How to Make Your PhD Work

How to Make Your PhD Work

A Guide for Creating a Career in Science and Engineering

Thomas R. Coughlin, PhD

Senior Director, Oncology Pharmaceutical Market Research and Strategy
Adjunct Professor of Innovation
Stevens Institute of Technology
Owner, PhDSource.com, NJ, USA

Published by John Wiley & Sons, Inc., Hoboken, New Jersey.
Published simultaneously in Canada.

For general information on our other products and services or for technical support, please contact our Customer Care Department within the United States at (800) 762-2974, outside the United States at (317) 572-3993 or fax (317) 572-4002.

Wiley also publishes its books in a variety of electronic formats. Some content that appears in print may not be available in electronic formats. For more information about Wiley products, visit our web site at www.wiley.com.

Library of Congress Cataloging-in-Publication Data
Names: Coughlin, Thomas R., author.
Title: How to make your PhD work: a guide for creating a career in science and engineering / Thomas R Coughlin.
Description: Hoboken, New Jersey: Wiley, [2024] | Includes index.
Identifiers: LCCN 2023032188 (print) | LCCN 2023032189 (ebook) | ISBN 9781394193141 (paperback) | ISBN 9781394193158 (adobe pdf) | ISBN 9781394193165 (epub)
Subjects: LCSH: Doctoral students–Vocational guidance. | Science–Vocational guidance. | Engineering–Vocational guidance.
Classification: LCC LB2386 .C67 2024 (print) | LCC LB2386 (ebook) | DDC 378.2/023–dc23/eng/20230718
LC record available at https://lccn.loc.gov/2023032188
LC ebook record available at https://lccn.loc.gov/2023032189

Cover Design: Wiley
Cover Image: © photoexpert117/Adobe Stock Photos

Set in 9.5/12.5pt STIXTwoText by Straive, Pondicherry, India
SKY10056804_100423

Dedication

As I interviewed more and more PhDs for this book, I learned that each one, like me, had to fight for their career. Nothing was handed to us on a silver platter. I listened to stories of absent advisors who abandoned PhDs to figure out how to do their own research. I heard other stories of advisors who lost their tenure amidst students' PhD programs. There were even stories of advisors who wanted to fabricate data and put pressure on their PhDs to publish it.

Separately from the interviews, I also conducted the literature search for this book and discovered countless articles enumerating reasons why the PhD system is broken. The articles explained that the programs needed to be fixed, yet offered few implementable solutions and lacked any semblance of responsibility. Granted, the problem is complex. Articles and stories detailed institutional problems that claimed the PhD is a hoax. They said entire PhD programs are farces because they lead students into underpaid PhD positions in advisors' labs where they are kept, for years, sometimes not making any progress in research and therefore stifling their careers.

But these experiences do not tell the whole story because the pure pursuit of research still exists within academia. The altruistic pursuit of the unknown has a silver lining somewhere. The pushing of oneself through low pay, uncertainty, and the lack of self-interest for the greater good still exists.

When I wrote this book, I chose to deal with the PhD system in its current form. I wanted to meet you where you are with your PhD or postdoc and guide you through your career transition out of the PhD by providing a lens on how to learn the data, understand the big picture, assess your situation, learn from other PhDs, and navigate your PhD career transition through this otherwise unclear and uncertain system.

This book will give you a realistic perspective on earning a PhD. If you follow the guidance of this book and complete the exercises throughout the text, you will have a more successful PhD or postdoc. The interview transitional stories provide key insight into the often rocky road taken by PhDs. Read them and learn from these PhDs' experiences. They are exceptionally honest and open stories.

With all the institutional obstacles, outside economic market dynamics, and changes in governmental funding allotments, there is still the singular PhD yearning for their own time and space to do research and hoping for a chance at a great job to demonstrate their worth. The pursuit of knowledge through pure research continues to exist in the hearts and minds of PhDs. There is still that silver lining. And there are plenty of careers that will fulfill PhDs and advance science. This book will show you the many avenues to success.

This book is for all the PhDs who deserve a fighting chance.

Contents

Preface

When I completed my PhD in 2015 from the University of Notre Dame in bioengineering, I believed I had gotten the most from my experience. I was looking back at five years, during which I had presented at conferences, received a fellowship, and published a few first author papers. I thought, "I must be competitive for a future career in academics." And while shaking the hands of my PhD committee members at my defense, I figured that I could someday stand in their shoes. I knew I needed a postdoc to gain grant writing experience to be competitive for an academic job, and I decided that was the best next step for me.

When I entered my postdoc, that logical reasoning and methodical decision-making gave way to crisis management. In the middle of my two-year postdoc fellowship my advisor was denied tenure and had to leave the university. I remember feeling lost, but I also believed my success should be independent of my advisor, so I kept moving forward. Unfortunately, as time went on, I realized I could not sustain myself on my modest postdoctoral NIH grant salary and small research budget. I needed proper training, and a second postdoc was the only way to get it.

I had heard about the postdoc crisis, and the trap that PhDs can get into of going from one postdoc to the next without any change in results. The idea of starting over in a new lab without any guarantee of a future academic career was overwhelming. I was all too aware of the financial burden of pursuing a postdoc and the economic, emotional, and professional costs of prolonging the transition to industry, if that was to be the end result. Two to three more years without certainty was a risk I was not willing to take. Even though I never imagined going into an industry job, I decided this was the direction I was going to move in. The idea of wasting more years in a postdoc was convincing enough for me.

During my postdoc, I had access to career training at New York University (NYU) Langone Medical Center. NYU had received the National Institutes of Health (NIH) Broadening Experience in Scientific Training (BEST) award, and this two-million-dollar grant was earmarked specifically to promote PhD and postdoctoral career training. For the remainder of my postdoc, I tried to take as many training courses as I could that this grant had created at NYU. I did not want to miss out on the opportunity to transition my career with this wealth of support.

As I took more classes, I began to see the big picture of where PhDs fit into the economy. I met more and more fulfilled and happy PhDs who have meaningful careers. And the more I headed away from academics, the less I began to look back at that environment.

Having now been in a nonacademic career for six years, I have found my vast scientific and engineering knowledge to be very useful in industry. Moreover, I have even seen validation of my thinking as more and more industry leaders suggest that business should be run with a scientist mindset. A mindset that a PhD can really bring to the table each and every day. Instead of a lack of fulfillment or a siloed job, I have found industry to be ever-changing and exciting to be a part of. Learning the way it works and how to perform in it has been a passion of mine as I went from starting my own company, teaching entrepreneurship, to consulting for a few startups, and eventually working my way back to industry.

The path that I have taken is similar to the road of many PhDs creating their PhD careers. In this book there are twelve first-person transition stories from PhDs from India, China, Greece, Israel, Italy, and the United States. Each interview deconstructs the interviewee's advisor, project, and environment, and then discusses how the PhD made their next career choices. Each PhD shares what they would tell a future PhD navigating this new landscape with different terrain than in the past.

This book contains the guidance, information, and empowerment that my peers and I wanted when we began our PhD and postdoc journeys in the late 2010s. To follow along with your PhD or postdoc, this book is broken into four parts:

- **Part I: You Are Here.** Understanding today's PhD job market.
- **Part II: Your Academic Path.** Navigating your PhD and postdoc and developing an academic career plan.
- **Part III: Your Nonacademic Path.** Comprehending nonacademic careers and how PhDs fit into careers in industry, government, and other economic sectors.
- **Part IV: Becoming the Proactive PhD.** Optimizing your PhD and postdoc to become a proactive PhD.

I intend this book to be a continued reference to you throughout your PhD or postdoc. Write in this book. Scribble on the PhD diagnostic tests to learn how your PhD or postdoc is going. Jot down your ideas in the career evaluation questions. Discover your perfect career path. And use the anecdotal evidence from the transition stories to help yourself and be inspired for your own career.

This book is for you, with all the knowledge many of my peers and I have learned. It is the expert advice, that as PhDs, we wish we'd had.

Best of luck,
Tom

Part I

You Are Here

1

The Twenty-First Century PhD

1.1 A Realistic Perspective

The story of my PhD is similar to the story of many of the tens of thousands of PhDs who graduate each year. During my graduate school experience, I did not receive nonacademic career training. Instead, I was taught how to become a principal investigator (PI). I learned how to run a research laboratory at a research-focused institution. Unfortunately, I was not taught how to convert my PhD into a job. And I was not alone.

The sad truth is that academic institutions do not budget enough money to support career training for PhDs (Malloy et al. 2021). And on top of that, your PIs do not have the in-depth knowledge of the job market to support your PhD career search (Dalgleish 2021).

Many PhD advisors want to help PhDs navigate their job search but just do not have the experiences to help them (Figure 1.1).

1.2 The Current PhD Landscape Has Changed

From the 1960s to the 1990s, the PhD was the primary gateway to a career in academics. These were the kind of erudite, wool-sweater-wearing academic professors you might have envisioned. You might have imagined yourself among them, strolling across the university quad, deep in contemplation, entering classrooms full of students captivated by your lectures. This epitome of a professor still exists, but the competition to get this job and to keep it has changed drastically.

After the early 2000s, the odds of turning a PhD into an academic career changed dramatically, shifting the career landscape for PhDs. Many factors have

How to Make Your PhD Work: A Guide for Creating a Career in Science and Engineering, First Edition. Thomas R. Coughlin.

Figure 1.1 Advisors might want to help you with your job search, but they often do not keep up with the jobs available for PhDs, and therefore, other resources, like this book, are needed.

contributed to the apparent decline in professorial positions, but the most significant factors influencing this have been:

1) the increased number of PhDs,
2) the decreased access to funding, and
3) the unchanged number of professor jobs.

1.2.1 Factor #1 : A Steady Rise in PhDs

One of the primary factors that has changed the job market for doctorates in the United States has been the steady rise in the supply of PhDs (National Science Foundation 2022). The number of PhDs granted in the United States has risen by approximately 3% on average since the late 1950s until today (Figure 1.2). In 2021, there were 52,250 doctorates awarded in total, and of those, 40,859 doctorates were awarded to PhDs focusing in the fields of science and engineering (S and E). This trending increase in the number of PhDs holds true globally, with doctorates increasing in countries throughout the world. PhDs being awarded doctorates increased by approximately 8% between 2013 and 2017 (Organisation for Economic Co-operation and Development 2019).

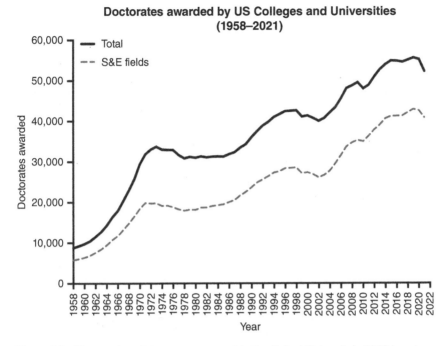

Figure 1.2 The number of doctorates granted in the United States since 1958 has risen steadily (National Science Foundation 2022).

1.2.2 Variable #2: The Funding Rates

One of the reasons that the professor job market changed is because funding has not kept pace with the growing number of PhDs. In the early 2000s, the success rate for grants was higher than it is now (Figure 1.3). For example, in the 2000s, the success rate was approximately 32%, but by 2022, it had dropped to 21% (National Institutes of Health 2023). This drastic change is due to more competition for funding. In 1995, there were 25,225 applications, and by 2022, there were 54,571 applications.

You might be wondering, what has the government done to increase this rate of funding? Well, they have increased budgets, but not as much as it appears.

For example, the NIH budget increased from $10.8 billion to $40.9 billion from 1995 to 2021 (National Institutes of Health 2023). (Figure 1.4) During this time, annual compound growth was approximately 5.4%. However, the average inflation rate was approximately 2.5% during this same time period (Statista 2023). As such, the increase in the NIH budget is not as substantial as it appears. If the US government just received a budget increase to account for annual inflation from 1995 to 2021, the budget in 2021 would be approximately $19.7 billion. Fortunately,

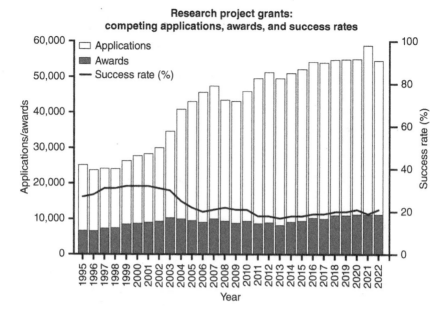

Figure 1.3 The NIH average success rate in the late 1990s and early 2000s was approximately 32% and has since settled at approximately 20% from 2006 to 2022. In 2022, the NIH success rate was 21%. Research project grants in this graph include DP1 through DP5, P01, PN1, PM1, R00, R01, R03, R15, R16, R21, R22, R23, R29, R33 through R37, R50, R55, R56, R61, RC1 through RC4, RF1, RL1, RL2, RL9, RM1, SI2, UA5, UC1 through UC4, UC7, UF1, UG3, UH2, UH3, UH5, UM1, UM2, U01, U19, U34, and U3R (graph type: stacked bar chart) (National Institutes of Health 2023).

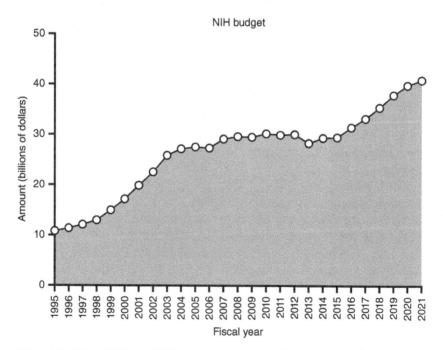

Figure 1.4 Since 1995, the NIH budget has grown from $10.8 billion to $40.9 billion in 2021, but the rate of inflation has been approximately 2.5% over this span of time. As such, accounting for inflation, the NIH budget increased from an adjusted $19.7 billion to $40.9 billion (National Institutes of Health Budget 2022).

the NIH budget has increased, but not as much as it appears. Therefore, the real increase in the budget from 1995 to 2022 is closer to two times than it was in 1995, not four times.

And comparing the number of people applying for grants between 1995 and 2022, there were 2.2 times more grant applications but an increase in the number of awards of only 1.4 times (National Institutes of Health 2023). So despite the fact that the number awards increased since 1995, as depicted in Figure 1.3, the number of applications has increased at a faster rate. This is one of the main factors accounting for the scarcity of grants and higher competition.

Let's consider the number of NIH R01 grants awarded in 2022. In 2022, there were 36,198 applications and 7,816 awards given at a success rate of 22% for the coveted R01 grant. By contrast, the success rate was 32% in 2000. So despite more money being available in 2022, the success rate is lower for R01 grants than it was 20 years ago.

The reason that the success rate has not increased with increasing funding available is because there are more professors applying and, therefore, more competition.

This conundrum stems from the rising number of doctorates being awarded each year. The graduate system in the United States simply does not set up PhD graduates to match with an academic career with a clear job path. In fact, the figures do not get better for academic jobs.

1.2.3 Variable #3: An Unchanged Academic Job Market

A report in 2013 in *Nature Biotechnology* on PhD careers and recruitment specified that there are 3000 new faculty positions in S and E fields created annually (Schillebeeckx et al. 2013). The number of available faculty jobs does not align with the number of new PhD graduates (Figure 1.5). PhD graduates are in abundance compared to the number of faculty positions, with there being a 10 : 1 ratio of PhDs to faculty jobs. This ratio however, is even lower due to the number of PhDs who stay in postdoctoral positions applying to faculty jobs each year. An earlier study reported that from 1982 to 2011, almost 800,000 PhDs were awarded in S and E fields, but only about 100,000 academic faculty positions were created in those fields over the same span of time (Schillebeeckx et al. 2013). Taken together, these numbers are not very promising for PhDs aspiring for academic professor positions.

Does this professor application success rate still hold true today? Let us test this with a job search. Running a job search during the hiring period for the fall semester of 2023 for full-time tenure-track professors of engineering and professors of science that required a doctorate returned 1,934 and 5,618 job results, respectively (Indeed Job Search 2023). Of course, the search algorithm also

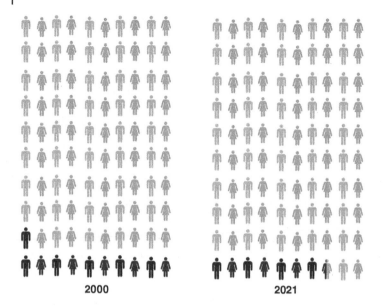

2000   2021

Figure 1.5 The percentage of PhDs who graduated in 2000 and 2021 were hired as professors in S and E, assuming there were 3000 professors hired each year (Schillebeeckx et al. 2013). There were 11 available positions per 100 PhDs who graduated in 2000 and 7.3 available positions for every 100 PhDs who graduated in 2021 in S and E. This estimation only accounts for PhDs who graduated in 2000 and 2021. It does not take into account the abundance of PhDs in postdocs who also might be applying for these job openings (National Science Foundation 2022).

included some positions that were not tenure track and slightly outside of S and E. Of the 40,857 newly graduated doctorates in S and E, 18% of these graduates could find jobs in the academic job market (National Science Foundation 2022). That leaves 82% of newly graduated PhDs to find other jobs. And this statistic is an underestimation because there would be more PhDs applying who graduated in previous years.

1.3   The PhD Job Market is Vast

So where do the majority of PhDs go if they do not go into academic careers? Despite the difficulties of trying to make it in academics, the current job market for PhDs is vast and genuinely exciting. In fact, I would argue it is more exciting than ever before.

More and more S and E PhDs are finding their way into nonacademic careers in government, nonprofit, industry, startups, the financial sector, and many other economic areas. Hiring managers have realized the potential of PhDs and appreciate the value they contribute to the workplace.

If PhDs are finding careers in nonacademics, then how are they getting there?

1.4 Conclusions

Up until now, there have been many PhDs creating their careers in academics and nonacademics without much guidance. The truth is, the competition in academics is so high that most academic advisors cannot give you advice on how to navigate the current academic landscape (McDonnell 2019). In addition, most advisors also cannot help you navigate the nonacademic career landscape. And although 90% of PhDs will go into nonacademic careers, the academic system trains PhDs to obtain jobs in research universities and not many other places (Cassuto and Weisbuch 2021).

Together with the knowledge in this book, there are 12 first-hand transition stories of PhDs after Part I, Part II, and Part III. These stories are from interviews with PhDs who actually navigated the career market in the last five to ten years.

Chapter 1 Key Takeaways
❖ The number of new PhDs on the market has steadily increased since the 1960s.   ❖ Academic institutions have not increased their rate of hiring full-time professor roles to match the increase in availability of PhDs.   ❖ Research funding has not grown at a rate to keep up with the rising number of PhDs.   ❖ With more PhDs on the market, PhDs have had to look for more alternative careers than the traditional academic career trajectory.

References

Cassuto, L. and Weisbuch, R. (2021). *The New PhD: How to Build a Better Graduate Education*. Johns Hopkins University Press.

Dalgleish, Melissa. (2021). Universities must do more to help Ph.D.s obtain skills for a variety of careers (opinion). *Inside Higher Ed* (26 August), https://www.insidehighered.com/advice/2021/08/26/universities-must-do-more-help-phds-obtain-skills-variety-careers-opinion (accessed 17 April 2023).

Indeed Job Search. (2023). Science and engineering PhD professor jobs. *Indeed: Job Search* (February 2023), http://indeed.com (accessed 11 March 2023).

Malloy, Jonathan, Lisa, Young, and Loleen, Berdahl. (2021). How the Ph.D. job crisis is built into the system and what can be done about it (opinion). *Inside Higher Ed* (21 June), https://www.insidehighered.com/advice/2021/06/22/how-phd-job-crisis-built-system-and-what-can-be-done-about-it-opinion (accessed 17 April 2023).

McDonnell, J.J. (2019). *Navigating an Academic Career: A Brief Guide for PhD Students, Postdocs, and New Faculty*. Wiley.

National Institute of Health. (2023). NIH Data Book – Success rates: R01-equivalent and research project grants. *NIH RePORTER*. https://report.nih.gov/nihdatabook/category/10 (accessed 11 March 2023).

National Institute of Health Budget. (2022). NIH Data Book – NIH budget history. *NIH RePORTER*. https://report.nih.gov/nihdatabook/category/1 (accessed 11 March 2023).

National Science Foundation. (2022). Doctorate recipients from U.S. Universities: 2021 | NSF – National Science Foundation. *National Center for Science and Engineering Statistics*. https://ncses.nsf.gov/pubs/nsf23300 (accessed 11 March 2023).

Organisation for Economic Co-operation and Development. (2019). OECD work on careers of doctorate holders. *OECD*. https://www.oecd.org/innovation/inno/careers-of-doctorate-holders.htm (accessed 11 March 2023).

Schillebeeckx, M., Brett, M., and Cory, L. (2013). The Missing Piece to Changing the University Culture. *Nature Biotechnology* 31 (10): 938–941.

Statista. (2023). Annual inflation rate U.S. 2022. *Statista*. https://www.statista.com/statistics/191077/inflation-rate-in-the-usa-since-1990 (accessed 11 March 2023).

2

The Real PhD Career Landscape

2.1 Your Advisor Cannot Be Your Only Guide

As mentioned in the previous chapter, you cannot solely rely on your advisor's advice to understand the PhD career landscape (Figure 2.1). Advisors lack the resources and experiences to show you all of the available PhD careers.

2.2 PhDs in Nonacademic Careers

PhDs go into a wide range of careers in both academic and nonacademic paths. However, these PhDs who have navigated the job market in the last 20 years did it without a road map. Oftentimes, these PhD's career transitions could have been more straightforward without so much turbulence, anxiety, and stress. Having an actual guide to these careers will help future PhDs not face these challenges.

The term nonacademic is a "catch-all" for any career outside of academics. Nonacademic career paths include careers in a multitude of fields, where PhDs, like yourself, are living happy and fulfilled lives.

Overall, PhD careers available to PhDs fall into several categories, including:

- Academics,
- Industry,
- Startups,
- Finance,
- Nonprofits,
- Government, and
- Technology transfer and intellectual property.

How to Make Your PhD Work: A Guide for Creating a Career in Science and Engineering,
First Edition. Thomas R. Coughlin.
© 2024 John Wiley & Sons, Inc. Published 2024 by John Wiley & Sons, Inc.

Figure 2.1 Advisors often lack the necessary resources and experiences to demonstrate all the potential career paths for someone with a PhD. You will need to use your own initiative and resourcefulness to perceive and understand the current PhD employment opportunities.

PhDs find jobs within all of these sectors, which span from marketing to equity analysts all the way to national scientists.

2.3   PhDs in Specific Careers

2.3.1   How Many PhDs are Going Into These Jobs?

In 2021, 72% of PhDs had definite commitments at graduation. These commitments included industry- or business-based jobs, academic positions, and postdoc positions. Only 36% of S and E PhDs ended up in academics, and looking at PhDs specifically in the physical sciences or in engineering, only 12% of PhDs had jobs in academics at graduation. However, this is not surprising, as evidenced by the numbers presented in Chapter 1. You might ask, if 72% of PhDs had commitments after graduation, then what about the other 28%?

Sadly, these 28% of PhD students were still looking for jobs after completing their PhDs. These PhDs might be working in their advisor's lab, working part-time as lab or teaching assistants, or, worse yet, unemployed.

2.3.2 A PhD, Unemployed?

I know how shocking this seems. PhDs all too often find themselves unemployed due to a lack of career training and support to help them navigate the career landscape. They simply do not learn the correct job statistics. Of course, that is until now.

Of the PhDs with commitments, they received a range of salaries that varied according to their chosen fields (Figure 2.2). For instance, S&E PhDs can earn

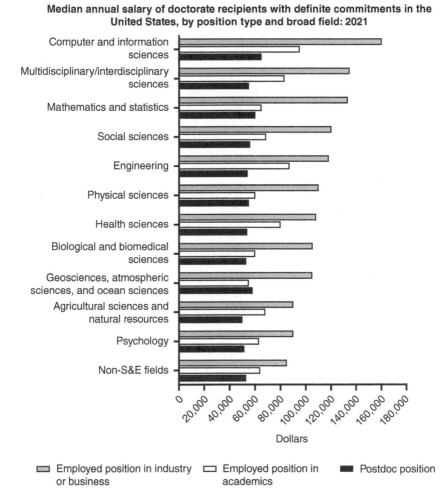

Median annual salary of doctorate recipients with definite commitments in the United States, by position type and broad field: 2021

Figure 2.2 Median annual salary of doctorate recipients with definite commitments in the US employed in industry or business, employed in academics, or in a postdoctoral position in 2021 (National Science Foundation 2022).

between $60,000 and $80,000 in academia and between $80,000 and $160,000 in industry. By contrast, postdoc salaries are low, with median salaries ranging from $50,000 to $55,000 in 2021.

The National Institute of Health (NIH) sets the postdoctoral salary at $54,835 for a postdoctoral fellow with zero years of postdoctoral experience (National Institute of Health 2022). By comparison, most entry-level jobs for a PhD have higher earning potential. The average salary of some PhD jobs in industry are:

- Scientist: $91,000,
- Medical Writer: $80,000,
- Biomedical Scientist: $100,000,
- Software Engineer: $112,000, and
- Project Manager: $80,000 (Payscale 2023).

There is quite a difference in earnings when comparing a postdoc to an industry position.

2.4 Self-assessment and Research

How do you decide what to do? As you go through exercises in this book, you will receive guidance on seeing if you are competitive in academics, perform a self-assessment to determine your next career, and finally, learn the steps to make that career happen.

As you move into the next chapter, you will get a realistic view of your PhD or postdoc. Then, in later chapters of Part II and Part III, you will gain information about careers in both academic and nonacademic careers.

2.5 Real PhD Career Transition Stories

As mentioned at the end of Chapter 1, the transition stories in this book span from the person's start of their PhD to their current career. They include individual lessons from real people navigating all types of PhD careers in various sectors of the economy. The stories are based on interviews with the individuals, written in first-person accounts that each person has approved.

2.5.1 PhDs With Widely Different Experiences

The featured PhDs have navigated the PhD job market within the last 10 years and pursued vastly different careers. There are stories from US citizens and from international visa holders. They include testimonials of how PhDs managed their job transitions, the lessons they learned, the challenges they faced, and struggles they overcame.

2.5.2 Diagnosis of Their PhD

The PhD transition stories break down their PhD experience into three components: their advisor, their environment, and their project. By breaking their PhD down you can easily compare your experience with the people's stories. This was intentionally done to show the importance of each component on a PhD's experience.

2.5.3 Stories Span Many Career Endpoints

The transition stories are from PhDs who are from PhDs in different stages of their career. There are stories from PhDs whose transitions are still in progress. Then, there are stories from PhDs who have successfully navigated academics. And finally, there are stories from PhDs who have moved into nonacademic careers, like government, industry, and finance.

Here is a breakdown of those stories and where they appear.

Transitions in Progress (after Part I)

- Vineeta Sharma, PhD – from India, postdoctoral fellow
- Sreemoyee Acharya PhD – from India, postdoctoral fellow
- Jesminara Khatun, PhD – from India, postdoctoral fellow

Transitions into Academics (after Part II)

- Antonio Marzio, PhD – from Italy, tenure-track Assistant Professor
- Ada Weinstock, PhD – from Israel, tenure-track Assistant Professor
- John Ruppert, PhD – from United States, tenure-track Assistant Professor

Transitions into Nonacademic Careers (after Part III)

- Leon "Jun" Tang, PhD – from China, Senior Director at a venture funding company
- Elizabeth Agadi, PhD – from United States, previous Graduate Career Services Consultant
- Laura Zheng, PhD – from United States, Data Scientist in the healthcare tech industry
- Amar Parvate, PhD – from India, Staff Scientist at a US national laboratory
- Henry Cham, PhD – from United States, Data Scientist in the Insurance industry
- Giannis Gidaris, PhD – from Greece, Property Treaty Underwriter in the reinsurance industry

You will gain the ability to do what many PhDs who preceded you did not have the opportunity to do. You will be able to learn early on the strengths and weaknesses of your PhD, learn from other PhDs, and take charge of your PhD with actionable items to bolster your resume and career readiness. All of this will leave you more at ease and aware of the opportunities available to PhDs and give you peace of mind that you have done your homework to give yourself the most information and advantage.

2.6 Conclusion

It is hard to imagine landing the career of your dreams when you are in the middle of the path. But trust me, you will get there. You will find the career you desire that allows and encourages you to continue growing and evolving. Many PhDs have set good examples for those entering the job market now. A PhD can command a level of respect in key roles that are relevant to their skill sets. You will read about other PhDs who triumphed over the most difficult of situations and found the empowerment to fulfill their career goals.

In the next chapter, you will learn about the PhD Career Feedback Loop. With this framework, you will become open to learning about your PhD and what you can do with it. Then, in Chapter 4, you will take the Diagnostics Test to obtain an objective understanding of how your PhD and postdoc are going in order to determine if you are competitive for an academic research professor position.

Chapter 2 Key Takeaways

❖ There are many satisfying careers for PhDs in industry.
❖ PhD-level jobs outside of academics have increased over time.
❖ Companies have learned the value that PhDs bring and pay them accordingly.
❖ Postdoc positions are low paid and necessary if they fit your career goals. However, postdocs can also be time losses if you do not stay disciplined.

References

National Institute of Health. (2022). How much is NIH's annual stipend for a postdoc with a fellowship award? *National Institute of Allergy and Infectious Diseases (NIAID)*. https://www.niaid.nih.gov/grants-contracts/annual-stipend-postdoc-fellowship-award (accessed 15 May 2023).

National Science Foundation. (2022). Doctorate recipients from U.S. Universities: 2021 | NSF – National Science Foundation. *National Center for Science and Engineering Statistics.* https://ncses.nsf.gov/pubs/nsf23300 (accessed 11 March 2023).

Payscale, Inc. (2023). "Doctorate (PhD) Salary." *PayScale.* https://www.payscale.com/research/US/Degree=Doctorate_(PhD)/Salary (accessed 27 April 2023).

3

The PhD Career Feedback Loop

"If you don't turn your life into a story, you just become a part of someone else's story."

– Sir Terry Pratchett, OBE

3.1   Deciding Your Own Story

According to fantasy author Sir Terry Pratchett, if you do not take steps toward your own story, you will be swept up into someone else's. The truly challenging part about forging your own path is that you do not always have a complete roadmap laid out for you. The good news is that you do not need the entire roadmap. You simply need to understand – and take – the next best step.

3.2   An Iterative Process

This chapter explores the "PhD Career Feedback Loop" and provides a practical approach for iterating on and searching for your niche in the doctoral job market. By leveraging strategies from both sides of the feedback loop, you will learn how to break through barriers and launch your career into a position perfectly suited for your strengths and aspirations.

3.3   PhD Career Feedback Loop

Deciding which door to walk through when you have a PhD can be intimidating, but it does not have to be. It all starts with self-reflection and ensuring that you know what you want so that you can make the best career decision. With the help

How to Make Your PhD Work: A Guide for Creating a Career in Science and Engineering,
First Edition. Thomas R. Coughlin.
© 2024 John Wiley & Sons, Inc. Published 2024 by John Wiley & Sons, Inc.

3.4 Sense and Respond | **19**

Figure 3.1 Balancing your self-assessment, feedback from your PhD diagnostics test, and your own personal goals with the PhD career research you conduct, you will move closer to the ideal fit for your future career.

of this book's self-assessment, PhD diagnostic test, and personal goals exercises, you will be able to determine your best career fit.

Whether you know it or not, you will be using principles for product design when searching for your ideal profession. The product design method involves bringing market research insights back to the design team to ensure the product meets the market needs of the customer. You are the customer in this situation, and you will conduct career research and compare those findings to your own personal interests and goals (Figure 3.1). This will help you find the perfect route to take on that career roadmap.

3.4 Sense and Respond

The ability to adapt to the current climate, market, or economic times is critical to having a business or research lab stay afloat and thrive. Becoming personally agile or nimble with your business or research is a style of management necessary in today's fast-changing world (Seiden and Gothelf 2017). Being able to evaluate your own PhD or postdoc through self-assessment empowers you to design your life. Using the diagnostics test in the next chapter, you will gain an objective perspective on your current situation through an objective lens. From there, you will acquire the tools you need to proactively pursue your goals.

3.5 Commit 100%

It is essential to wholeheartedly dedicate yourself to the path you are pursuing. If an academic PhD career is what you desire, then make sure to give it your all. Equally, if you decide to leave academics and take a nonacademic route with your PhD, it is important to maintain your job search as a top priority. When we are fully committed to something, incredible results can be achieved through the growth experienced in that process. This direct investment in our goals comes with many advantages, such as gaining credibility and respect, realizing personal progression and satisfaction, as well as increasing efficiency due to effective time management. Therefore, dedicating 100% of ourselves should never be taken lightly; it could mean the difference between success and failure.

3.6 Conclusion

To summarize, deciding your next career step is a key component of getting through your PhD or postdoc. Taking the steps to apply the PhD Career Feedback Loop will ensure that you are making conscious decisions about your future and that you are able to iterate, as needed. It is important to remember that you have the power to decide for yourself what works best for your career and that no one else can make those decisions for you. Put some trust in yourself and try out the steps outlined today. If anything needs adjusting or revising, it is important to keep track of what worked and what did not in order to continuously improve upon your chosen path. Getting through a PhD takes a lot of work, but you have taken the first step toward success – taking action!

Chapter 3 Key Takeaways

- ❖ During your PhD job search, you can use the PhD Career Feedback Loop to perform self-assessment, analyze your diagnostic test, and reflect on your personal goals while simultaneously looking for careers in academics and nonacademics to find your "Career Fit."
- ❖ The mindset needed for the PhD job search is that you need to commit 100% to the career path you choose.
- ❖ By committing 100% to your career choice, you can give it your all.

Reference

Seiden, J. and Gothelf, J. (2017). *Sense & Respond: How Successful Organizations Listen to Customers and Create New Products Continuously.* Harvard Business Review Press.

4

An Objective Assessment of Your PhD or Postdoc

4.1 Same Three Letters With Very Different Experiences

The process by which PhDs obtain these three letters following our names is neither simple nor straightforward. These three letters do not even begin to capture
the essence of your PhD experience. The PhD demands our energy, time, sweat,
and tears. It also requires the kind of determination, dedication, willpower, self-
discipline, endurance, and resilience that many people just do not have. And
while those three letters may not encapsulate or convey the true breadth of your
PhD experience to someone without a PhD, they will inspire respect and esteem
from others throughout your life.

The uniqueness of our PhDs is defined by our advisor, project, and environment. In this chapter, you will examine the impact of your advisor, your project,
and the environment on your PhD or postdoc. Then, you will take the diagnostic
test to obtain an objective assessment of your PhD or postdoc.

4.2 Taking the First Step

Afraid of taking the first step toward analyzing and assessing your PhD or postdoc? Don't be (Figure 4.1). You need to see what it is all about in order to know
how it works. The better you understand what can go wrong, the better decisions
you can make to avoid those pitfalls.

How to Make Your PhD Work: A Guide for Creating a Career in Science and Engineering,
First Edition. Thomas R. Coughlin.
© 2024 John Wiley & Sons, Inc. Published 2024 by John Wiley & Sons, Inc.

Figure 4.1 Taking an objective look at your PhD or postdoc with the diagnostic test is the first step toward taking ownership of your PhD.

4.3 The PhD and Postdoc House

Having a good PhD and postdoc is like constructing a house. It is important to have a strong foundation and well-made scaffolding for you to move within to furnish and live in this house. A poorly made house with a bad foundation will easily fall down when you start living in it (Figure 4.2).

Your advisor forms the foundation of your PhD and postdoc house, while the environment they create and the projects they work on are the scaffolding. With a strong foundation, it is easier to put well-made scaffolding in place. Overall, a PhD hinges on having an advisor who is a good leader. Having this foundation will make your PhD or postdoc a very enjoyable experience for you to feel supported.

4.4 Advisor, Environment, and Project, in That Order

Following the completion of my postdoc in 2017, I researched the factors that contributed to both positive and negative PhD experiences. I interviewed many PhDs and postdocs during my research whose experiences with their advisors

(a)   (b)

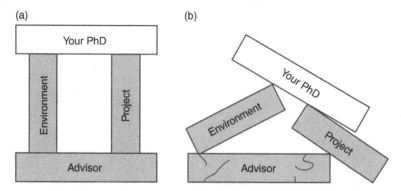

Figure 4.2 (a) The success of a PhD and postdoc depends on a strong foundation, which is created by having a good advisor. A good advisor will provide a good environment and help foster a strong project. These will allow for a positive experience in your PhD. (b) By contrast, having an advisor who is not supportive or not present will not enable a great environment or foster the ability for you to have as successful of a project.

ranged from dreadful to delightful. As a result, I discovered that the advisor was the single most significant aspect of a PhD.

Those who endured unpleasant and unsatisfactory PhD experiences were typically limited mostly by their advisors, then their environments, and finally by their projects. They frequently wished they had placed more emphasis on their interactions with their advisors when interviewing at the start of their PhD. Not every PhD had an awful experience with their advisor, but many who overlooked their ability to "get along" with them expressed regret.

On the opposite end of the spectrum, PhDs and postdocs who shared accounts of extremely joyful experiences spoke of healthy interactions with their advisors. The happiest PhDs and postdocs had advisors who were dependable, attentive, and consistent.

Your advisor's influence on your PhD experience can shape your perception of academics and often color the career options that you might pursue. If you have a choice, it is worth the extra effort to apply to more universities – or even wait a year to apply – in order to ensure that you have a great advisor during your PhD. Finding an excellent advisor who is also a good fit for you is absolutely essential.

The postdoc advisor is equally important in fostering a supportive environment. At this point in your career, you are more focused on the job market. Having a supportive advisor will give you the boost you need to ensure you reach your goals. That said, the PhD advisor is the most important, and choosing one you "get along" with is crucial.

4.5   Afraid of What You Might Find?

It is okay to not want to look at the objective assessment of your PhD. It is okay to realize that your situation could be better or you are stuck in a less than great postdoc or PhD.

But no matter if you are in a less than great PhD or postdoc, this book will help bridge the gap to help you navigate through your PhD and learn about obtaining the right career.

4.6   Checking How Things Are Going

Throughout your PhD and postdoc appointment, it is vital to check the pulse of your progress. At some point in the process, you will get the question from peers, friends, or family, of "How's it all going?" In most cases, many of your friends and family will not understand what it takes to do a PhD or postdoc. Many of your friends from secondary school or college might be in "industry" and they will not truly understand the way your path in academics works. Similarly, you might not know how to tell them how it is going without knowing what the PhD or postdoc should be like.

Therefore, to answer the question of "how's it all going?" you will first have to know where you are going and then, know if what you are currently doing is getting you there. So let us talk about how to know, "How's it all going?" and get an objective assessment of your PhD or postdoc.

The PhD or postdoc can be broken down into three main factors:

1) Your advisor,
2) Your project, and
3) Your environment.

The self-assessment diagnostic test is specifically designed to give you an objective lens of your PhD or postdoc by assessing three factors: your advisor, your environment, and your project.

4.6.1   What Do With the Results From the Test

Overall, understanding each factor of your PhD or postdoc will empower you to proactively solve or mitigate specific situations that you might be experiencing. In the following figure, we depict a quick analysis of what "The Right Stuff" and "The Wrong Stuff" are for these three main factors (Figure 4.3). Take a look to see what you have and then continue below for an in-depth understanding.

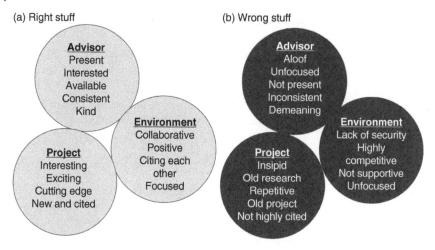

Figure 4.3 Your PhD or postdoc can be made up of a mixture of the right and wrong stuff. (a) Right Stuff. Having a great experience can really come down to having the right stuff, and vice versa. Your PI, or advisor, is the most critical component and really sets the tone for the entire experience. Advisors with characteristics like being present, interested, available, consistent, and kind can set the tone for the entire program and group, fostering a positive environment. The project is directly influenced by the advisor and environment, and when it is supported with consistent meetings or oversight, you will be able to thrive. Of course, as a secondary importance after the advisor, it is important to pick a project that is interesting to you and will keep you excited for the duration of your PhD or postdoc. (b) Wrong Stuff. By contrast, the wrong stuff can be destructive in many ways, manifesting in anxiety, depression, and a lack of progress. Identifying your PhD and postdoc factors early on can help you ameliorate certain parts of your program or show you whether you should move on.

4.7 Factor #1: Your Advisor

Your advisor is the single largest influence during your PhD experience. PhDs sometimes speak of low morale and the feeling that their research is crawling toward a dead end. In these situations, the advisor was so immersed in getting and maintaining their grant status that they had little time to look into research or advise students. In one example, a student spoke of spending two years working on a stem cell separation project without ever receiving feedback on their lack of progress. In the end, this student received their PhD and succeeded in an industry career.

If your advisor is good at getting grants, good at maintaining structure and a schedule in the lab, and good at communication, they will be able to shield students from some of the obstacles discussed. Yet, keep in mind, having an advisor who cares is really the key component. Your experience should not mirror a "puppy mill," where degrees are pumped out without care for learning. Nor should it be one that sets students against students or has you working 70 hours a week. A caring advisor will discuss your future as if it is their own.

4.7.1 The Importance of Your Advisor

Out of all the factors influencing your entire PhD program, the most important is your PhD advisor. PhDs, including the case studies in this book, discuss the importance of choosing an advisor. In all of the case studies of PhDs reflecting back on their PhD experience, they expressed that having an engaged and interested advisor was the most important part of their degree and impacted their happiness the most. In all cases, PhDs indicated that they would rather work for an advisor who was kind and good-natured on an uninteresting project than for an advisor who was well-funded but lacked transparency and morality.

What makes a good advisor? A caring, consistent, interested, and present advisor are key characteristics of a good advisor. By contrast, you might ask, what makes a bad advisor? PhDs who report having bad advisors enumerate that the advisor is defined by a lack of focus, empathy, compassion, attentiveness, and presence. In other words, there was a lack of strong leadership within these labs or groups. And overall, the lack of leadership stemmed from the advisor.

Your success will largely be dependent on your advisor. Knowing this will help you determine what PhD and postdoc program will be beneficial for you, and in Chapter 5 I list ways to improve your relationship with your advisor.

4.7.2 Hallmarks of a Good Advisor

Your advisor's attitude will reflect directly on the lab's attitude. The advisor's ability to see their students as potential collaborators rather than only workers will enable a positive environment. Similarly, the advisor's ability to empower students to do research will help them become scientists, engineers, and/or self-starters. Empowered students are happy ones.

4.7.3 Hallmarks of a Bad Advisor

Obviously, the opposite of an empowering mentor is a controlling, disenfranchising advisor. These advisors might be so busy that they only care about outcomes; they might be under so much pressure that they look over your shoulder; or they might be distrustful of your work and subject you to a "search and frisk," where you and your work are analyzed and micromanaged. However, keep in mind that this situation can be better than the "hands off" approach, where students find themselves at the coffee bar four hours a day and then out to lunch.

Going one step further, there are situations where some advisors are not meant to be PhD advisors. These advisors are disorganized, out of touch with reality, and poor communicators. In my interviews with PhD students, some of the worst situations that PhDs experienced were:

- Advisors who would yell and criticize you,
- Advisors who forced you to publish research before it was finished,
- Advisors who promoted an environment of fabricating data, and
- Advisors who ignored you and your progress.

4.7.4 Advisors Who Lose Tenure

Some PhDs and postdoctoral fellows will have an advisor who goes up for tenure review and does not get it. When your advisor is nontenured, there is always the possibility of them losing or not achieving the benchmarks for tenure during their review. Although nontenured professors on the tenure track obtain feedback from their tenure review committee during tenure review committees that happen at specific points along their tenure track, there is always the chance that the advisor might not receive tenure.

When an advisor does not make tenure, they might not be a bad advisor. In most cases, these advisors are perfectly qualified but did not bring in enough funding for the university. Because of this, the advisor is asked to leave.

If you are in an advisor's lab or group that does tenure, you might need to find a new advisor, wrap up your PhD in their absence, or move to a new position at a new school with them. There is never a perfect solution, but depending on where you are at, you might find one of these options better for your life and goals.

4.8 Factor #2: Your Environment

The environment consists of several factors: your university, advisor, lab members, funding to the lab, lab capabilities and equipment, and the general feel or atmosphere of the different labs or facilities. The root of what makes a good or bad environment is dependent on multiple factors:

- Does the advisor have tenure?
- How much pressure is the advisor under from the University?

These are two questions that can be helpful in determining the situation you are in.

4.8.1 Impact of Your Institution and Advisor on the Environment

High pressure from the university on advisors to produce grants and generate money for the university equals stressed-out researchers under pressure to produce. Similarly, if the advisor does not have tenure and the tenure necessities of the University are very high, the professor will likely be under a great deal of pressure to produce quality publications, obtain grant funding, and build a strong research profile. Of course, how the advisor handles the pressure is also important, and the strength of their research and leadership is critical to success. If the advisor is more latent in their career or not as motivated to do research, this will affect the general environment.

4.8.2 Primary Research Institutions or Medical Institutions

Certain universities are under greater restrictions to produce research. In medical universities, the professors' pay can come 50% from grants and 50% from the academic institution (Bloch 2010). Professors at these institutions are often required to obtain grants in order to hold their job position. As such, tenure needs can be very high for these professors because more is expected of them from their research.

4.8.3 Research and Teaching Institutions

Pursuing a PhD at a university that is invested in both education and research can be a very positive experience. To know if you are in a research and teaching university, the professors will have expectations of both teaching and conducting research. In academic institutions, the university can fund the professors at 75% and have 25% of their salaries come from overhead on grants. This can create a more stable position for the professor as typically, these institutions' professors are not under as much pressure to produce grants or research, because they are expected to teach.

These environments put more focus on different aspects of the professor's production capabilities. They are either asked to produce more research and their job depends more on this (medical institution), or they are asked to focus more on teaching and research, enabling a more academic environment. Both environments are good. Medical institutions can be more focused on translating basic science into clinical products and offer different insights and experiences than an academic institutions can.

4.8.4 Research Group Members

The advisor sets the tone of the group. A toxic environment can be created by unclear guidance from an advisor, causing uncertainty within the group or lab.

Without clear leadership and guidance, researchers may feel anxious and lack the positive reinforcement they need to explore their creativity and ideas. This can manifest into many problems, including some pretty nasty extremes. Awareness of the fact that you are in a toxic environment can be understood over time and by also comparing another person's situation to your own.

4.8.5 Nontenured Advisors

Obtaining tenure is a difficult process that differs from institution to institution. The work to obtain tenure can be insurmountable and in some cases it actually is insurmountable.

Working in a lab where a professor is seeking tenure has upsides and downsides. Nontenured professors offer many positives. They are usually hard working. With that, they are usually very reliant on a PhD student or postdoc. They will be very motivated to build a great research program, and a position in a nontenured faculty lab will give them a breadth of experience unmatched in other positions.

PhD candidates should recognize that there are always unknowns lurking in their programs. A professor's opportunity to obtain tenure could be ended, forcing them to relocate to another university or possibly leave the country for another opportunity. You may need to move with them, as funding could have been transferred. This could mean starting over in a new lab in order to remain at the institution.

Other unforeseen drawbacks might include a trickle-down of stress originating from high tenure demands of the department or your professor. Usually, this type of stress can be managed with open communication between lab partners and your support system. If the stress is unmanageable, you may have to make a choice, put your nose to the grindstone and do good, consistent research, or leave the position. Sometimes, difficult circumstances can greatly affect all aspects of your life. If you are in a difficult situation, all you can do is do good research and be consistent. For example, in a lab where the hourly week climbed from 45 hours to 60 hours, plus weekends, the professor was pleased with the amount of work, but the lab's atmosphere became toxic. Students were in stiff competition to produce papers and get their names out there in the research community. Only the most rigid students were able to keep this schedule. For the rest, it was a recipe for burnout. In an ill-managed lab, a professor's approach and attitude in this situation can really make or break the environment. They should be working harder than anyone else and wanting to succeed more than anyone else.

4.9 Factor #3: Your Project

4.9.1 Determinants of a Good Project

The project you will work on will stem from the situation you are in. Funding for a project will be the main criteria for success of the project. Throughout the lifetime of a PhD or postdoc, the professor's success will largely depend on their funding. New nontenured professors are often given generous start-up packages to construct research programs, so they will be capable of carrying out research without government or outside funding, but eventually will need to solidify some source of funding for their research (Bloch 2010).

Having a funded project will give it direction and enable a clear publication plan. As the best "laid plans of mice and men often go awry," the research will always take unexpected turns, even on a funded path, but a funded project will enable more stability.

A nontenured professor may not have grants yet and is more than likely still carving out a research niche and working to produce preliminary data for grants. Sometimes, the first task for these PhD students or postdocs will be to write a review paper in order to identify themselves in the new field and start to garner a public association within a given research field.

4.9.2 An Inherited Project

There is another type of project – the inherited project. This is a project that as you started your PhD, you were given a project to *finish* from a previous graduate student. These can be huge positives if it is looked at as such, but due diligence should be done to finish this project and move onto something new or identify a new direction to spin off from the project. Oftentimes, in this scenario, you might not be the first author of this project.

The uninteresting project; this project is in a direction that seems exhausted and is not affecting the world. These projects should be avoided at all costs. This can lead to depression and be a difficult situation to be in. The research for your PhD should be exciting.

4.10 PhD and Postdoctoral Self-Assessment Diagnostic Tests

Take this test at the end of your first year and learn how your PhD is going from an objective point of view. There are more tests in the Appendix.

1. Advisor (Supervision, Mentorship, Support)

1a. Does your advisor have weekly meetings with you or the group?	Yes (0)/No (1)
1b. Does your advisor give you advice on how to start research projects?	Yes (0)/No (1)
1c. Is your advisor encouraging?	Yes (0)/No (1)
1d. Does your advisor introduce you to people at meetings, conferences, or within your institution?	Yes (0)/No (1)
1e. Does your advisor frequently check-in on you at your desk or in your lab?	Yes (1)/No (0)
1f. Is your advisor present there (within reason) when you need them?	Yes (0)/No (1)
Total	_____

2. Project (Funding, Type of Project)

2a. Is your project a continuation of another project?	Yes (1)/No (0)
2b. Is your project something that your advisor or group needs to finish and move on from?	Yes (1)/No (0)
2c. Is your project new and novel?	Yes (0)/No (1)
2d. Is your project considered exciting in the field?	Yes (0)/No (1)
2e. Is your project dependent on someone else's work?	Yes (1)/No (0)
2f. Is your project on the cutting edge of the field?	Yes (0)/No (1)
Total	_____

3. Environment (Group, Competitiveness, Institution)

3a. On average, are people in your group mostly available for help on a weekly basis?	Yes (0)/No (1)
3b. Do people in your group share information freely?	Yes (0)/No (1)
3c. Do the senior members in your advisor's group like coming into work?	Yes (0)/No (1)
3d. For new PhDs, do the people in your group seem genuine?	Yes (0)/No (1)
3e. Do other internal research groups compete with your group? Is the environment competitive?	Yes (1)/No (0)
3f. Are the existing systems or protocols in place to make the work orderly and efficient?	Yes (0)/No (1)
Total	_____

Tally Your Scores and Write Them Below.

1) Advisor   _____
2) Project   _____
3) Environment   _____

Scoring Rubric:

Good advisor, project, and/or environment	(score = 0–1)
Moderate advisor, project, and/or environment	(score = 2–3)
Poor advisor, project, and/or environment	(score = 4–5)
Terrible advisor, project, and/or environment	(score = 6)

4.10.1 Interpreting Your Scores

As you tally your scores, notice some of the metrics that had a 1 point associated with them. These are negative scores. So the higher the score you have, the more challenging of a situation you are in. In order to analyze your scores, you can see that projects, advisors, and/or environments each receive their own scores. Each one of these factors contributes to your overall success and happiness in your PhD.

4.10.2 What Does a Good Rating of 0–1 Mean?

A score of 0 or 1 informs you that you have a positive part of your PhD or postdoc. Perhaps you have two or even all three factors as being good. This is great. This means that your project is strong, your advisor is helpful and mentoring you as appropriate, and your environment is a value add and positive one. Having all three factors being good is not only a healthy, happy workplace for you to succeed in, but it is also a workplace that could set you up for a successful academic career if that is what you desire.

Even if only one of the factors is rated as good, you can lean into this one factor and try to focus on the positives of the other parts of your PhD or postdoc. This way you can optimize the PhD or postdoc experience that you have.

4.10.3 What Does an Intermediate Rating of 2–3 Mean?

A moderate rating is very good and even a PhD or postdoc with all moderate scores for each of the advisor, project, or environment can be a positive and strong experience. The point here is to focus on the parts of your PhD that are good and do a few things to make them better. By reading this chapter, you are already making the steps toward improving your PhD or postdoc and your overall experience!

4.10.4 What Does a Poor Rating of 4–5 Mean?

The truth is that having a 4 or 5 means that out of the 6 parts of each factor – the project, advisor and environment – you have most of the parts of them as being negative and unsupportive of your success. Of course a poor environment but a great advisor and great project can be a positive experience, but might lead you to seek out more helpful resources outside of your research group or institution in order to be able to provide a positive environment for you to work in!

4.10.5   What About a Terrible Rating of 6?

Yikes! Well, yes, this is not great. Most or all of the 6 parts of each factor are negative. This is not good, but not insurmountable. The hope of progressing through your PhD or postdoc will be a rocky road. It is going to require extra effort to make the experience positive. Furthermore, hopes of an academic job are now going to be a bit of a pipe dream. Of course, if only your environment is terrible, you might be able to trudge through the work and still have a great experience.

4.11   Conclusion

Congratulations! If you have taken this diagnostic test, you are much closer to understanding your PhD program and getting an objective assessment of your PhD. One of the most pivotal outcomes of this diagnostic test can be to help tell you if you have a competitive PhD for an academic position and whether it will be worth continuing strongly on the academic research professor route. Furthermore, if you are not going to be competitive based on the results of the diagnostic test, this does not mean there are no options. Instead, this result actually opens a plethora of options. Of which, these options may open up the most rewarding career option for you.

The PhD and postdoc experience can be some of the greatest and most full of learning years of your life. You cannot just walk into an experience expecting it to be great. You have to do your best to make it that way.

Chapter 4 Key Takeaways

❖ Just having a PhD by itself does not make you competitive for an academic career.
❖ During your PhD, you must be proactive within your environment to make the PhD more favorable and better for yourself.
❖ The environment, advisor, and project each impact how successful your PhD and postdoc will be.
❖ A negative environment or toxic laboratory can be an unaccounted for obstacle impacting your academic career path.

❖ An inexperienced or aloof advisor can disrupt the best PhDs and lead them to squander years of productivity.

❖ A recycled project or old or too radical of a research idea could be deemed unpublishable or uninteresting by the research community.

❖ Using the diagnostic test, you can objectively gain an understanding of how your PhD is going. There are more diagnostic tests in the Appendix for you to conduct additional assessments of your PhD or postdoc.

❖ The diagnostic test will give an insight into how competitive your PhD will be when applying for jobs.

❖ Overall, the diagnostic test will help you compare your advisor, environment, and project to all PhDs to tell you how competitive your PhD will be in your academic career plan.

Reference

Bloch RJ. (2010). Starting off as a tenure-track assistant professor in a school of medicine. [PowerPoint slides]. National Institutes of Health (.gov). http://www.training.nih.gov/_assets/slides_3_23_10 (accessed 20 May 2023).

Transition Story: Vineeta Sharma, PhD

PhD: Indian Institute of Technology Madras
Field of Study: Tissue Engineering
Current career position: Postdoc Researcher at Harvard University

Motivation for pursuing a PhD: After my Master's degree in Tissue Engineering from University of Calicut, Kerala, I decided that I wanted to be a professor. I thought that I could teach and carry out some of the tasks of being a professor and all of it seemed feasible. In my Master's program, many people advanced to pursue PhDs. This was a common path.

To apply for the PhD in India, you have to take a National Fellowship exam. I first did this and then applied for four PhD programs focusing around research in biomaterials. I was selected for three and decided to go to Indian Institute of Technology Madras (IIT Madras).

In the first year at IIT Madras, I took courses and passed my first-year viva examination. The viva was an oral examination of foundational knowledge I learned in my PhD coursework. If you do not pass your comprehensive viva exam, you typically have to leave. At IIT Madras, about 75% of PhDs pass the exam.

After my examinations, I rotated through three labs and selected the third lab because I liked the area of research, which focused on cardiac tissue engineering.

Comparing the Indian system with the US PhD system: In India, you first apply to the National Fellowship program and then if you are selected you are given a fellowship and then you interview at universities. However, in the United States, you typically need to apply to a PhD program and then already be in a PhD program to apply for a fellowship. In India, once you have the fellowship you know you have a stipend and funding to do some research. That said, similar to

How to Make Your PhD Work: A Guide for Creating a Career in Science and Engineering, First Edition. Thomas R. Coughlin.

the United States, the principal investigator's (PI) or professors still need to apply for grants to support their research funding.

In India, the competition for professor jobs can be very challenging. Many of the newly minted PhDs need to differentiate themselves to obtain a tenure track faculty position in an Indian university. Many PhDs do this by doing research outside of India in a postdoc position or two in Europe or the United States. By doing this experience, many PhDs are more competitive when applying for Indian professor positions. However, if you pass 35 years of age, you are not able to apply for a professor position and therefore, have aged out of the timeframe of being allowed to apply.

Family has a large effect on whether a person might return to India to try for a professor job. In most circumstances, a person will return to India if their family wants them to, and in other scenarios, a person might stay where they are if their family is supportive of that.

Advisor: My thesis advisor was very supportive of me and my work. My thesis advisor was really good and very attentive to me. He gave me the opportunity to explore my ideas and follow my discoveries. He helped me pursue my research interests and define these areas.

Environment: My PhD lab was big. We had twenty people and had weekly meetings every Friday. My labmates were simultaneously supportive of one another and competitive at the same time. We used to compete with one another to do more work, but it was not in a toxic way. We all wanted each other to get their work done and achieve their results.

Project: My project had an *in vivo* component that my PhD lab did not have and during my PhD I partnered with another lab that had in vivo research facilities.

Benchmarks: In India, you have to have two first author papers to submit your thesis, with one paper being published and the other being published or communicated. One of my papers got accepted before submitting my thesis, but the second paper took two and half years to get published. This second paper finally got published one year after my PhD defense. We had initially aimed for high-end journals and then tried to lower our sights to other journals and we were finally published.

Next career steps: During my PhD, I wanted to attend conferences to meet PIs in the United States to get a postdoctoral position. At two of the international conferences I attended, I networked and tried to find positions that would be open when I was finished with my PhD. It was difficult to find potential future positions that would sponsor a Visa application. Then, my PhD advisor notified me that Harvard

University was looking for a postdoctoral fellow. The lab was perfectly aligned with my interests and I was able to interview and obtain this position.

Postdoc position: My postdoc position is going great. My lab is a very nice and really fun place to work. My research is in kidney tissue engineering and is a very difficult topic, but my (principal investigator) PI knows it is a difficult topic and is empathetic to this. I am three years into my postdoc and at present and my PI asked me to stay a little longer.

In India, there is hierarchy in labs. In the United States, there is no hierarchy. In the United States the dynamics are very different than in India. I have more independence.

Biggest challenge: I have not had a big challenge. I expected getting my Visa for my postdoc in the United States to be my biggest challenge, but I have been really lucky. I got my appointment for the postdoc in May 2020 and then got my Visa by July. My postdoc PI was helpful and got the letter needed to get the Visa finished on time. For my Emergency Visa, I needed a letter from my PI saying why it was an emergency. He did a great job and the letter was strong. The government approved it right away. I have heard that sometimes the Emergency Visas get declined because the letter from the PI is not strong enough.

Career after postdoc: At first, I wanted to be an instructor at Harvard doing a tenure track, but now I am not sure what I want to do. The cost of living in Boston is really, really high and the postdoc salary is challenging. I am figuring out if I want a tenure track or industry position. My PI has recently offered me a third option of staying in his lab as a staff scientist. I am still debating.

Proactive steps: I am talking to a lot of people who shifted from academics to industry. I am attending lots of career development seminars at Harvard University to learn about that. For me, I am not deterred by having to move for a future tenure track position.

One consideration for me is that to get a green card, you can apply for the Employment-Based Immigration, first preference EB-1 Visa. This is a very exceptional green card. In industry, the challenge is more difficult. There are lots to weigh in, and I am still undecided and exploring the vast opportunities.

Mental health: A lot of people cannot handle the mental pressure that comes with the PhD. I have seen this first hand. In contrast to undergraduate coursework where you know the expectations, the PhD is open-ended and the solutions are not always there. You have to be able to handle stress really well and be okay with the unknown. You have to be okay with knowing that you just did your best to do the job.

Advice for future PhDs: For someone considering a PhD, I would say that a PhD is a lot of struggle. If you cannot fully invest yourself; if you are 10% not sure, then do not go for it. If you are not passionate about it, then do not do it. There will be challenging moments that you will need passion to get through. The PhD does not pay much. You have to be willing to work hard and sometimes put in 70-hour weeks. There is a lot of sacrifice and it can affect your mental health.

For someone considering a PhD who lives in India: For a PhD in India, I would say that there is a lot more supply and less demand. The number of people going for PhDs has increased and there are not a lot of jobs after just a Master's degree. There are lots of PhDs and not enough professors and oftentimes there are 100 people competing for a single job. In the United States, it is not as competitive.

Perceptions of my PhD from India: I have found that my PhD earned in India is viewed the same as one earned in other countries. I do not see any difference between European, United States, and Indian degrees. I find that most times it is how well you know the techniques and how well are you able to adapt to learning a new technique. Research is very respected in India. In the second paper from my PhD, I wanted to get it published in the Journal of Biomaterials (a high-impact factor journal for the field of tissue engineering). I have seen people with less work get published with a few big names on the paper in that journal. But for me, when your PI is less well known it can be difficult. Perhaps this was the challenge I faced. There are well-known PIs in India, so this was not based on my geographical location but rather on the esteem of my lab.

Key Insights:

- Taking steps to learn the options available to PhDs helps widen the lens so you can really see the full spectrum of jobs available to you.
- Handling stress through seeking support, meditation, exercise, or other means is vital to maintaining your mental health during your PhD.
- Doing a PhD might involve fully investing yourself and there are times when you will have to work hard at maintaining a balance in your life.

Transition Story: Sreemoyee Acharya, PhD

PhD: Iowa State University
Field of study: Microbiology
First career position: postdoc at Tufts University
Current career position: postdoc at Cedars-Sinai Medical Center

Motivation for pursuing a PhD: The reason I started thinking about a career in science was that my undergraduate professors encouraged me. In undergraduate college at Mumbai University, I had some really amazing professors. In India, you either, typically, go into medicine or engineering. These are the two broad classifications, and every single kid is conditioned to go into either engineering or medicine. And that is the funny thing, I did not want to do either. I wanted to go into something which was totally different.

Transitioning from India to the United States: I went into pure sciences and I genuinely enjoyed microbiology. I really liked the entire research aspect of it. I followed my undergraduate degree at University of Mumbai with a Master's at University of Mumbai. Again, I liked where my education was leading me. After graduation, the entire idea of the PhD was a little intimidating, so I decided, I would just try to earn a Master's Degree at the University of Central Florida. I wanted to see if I could transition from the education system in India to the education system in the United States.

I was given the option to pursue my PhD at University of Central Florida, but after two years there I wanted a different perspective. I wanted to go to a better institution. I also wanted to leave Florida. That is when I decided I would go into Iowa State.

I personally think that I was at a good spot when I was applying for a PhD because I was already in the United States and I already had a good GRE score. I had good transcripts. Overall, I think it was not that hard for me to get a PhD position. The applications and getting into the program were not that hard because I had already been in the United States for two years and I knew how to make statements and how to send the proper emails. I also knew my research interests.

Starting the PhD: I received the full perspective of the academic system. From the outside, it looked all rosy, but I went through so many bumps and hassles. It is not as easy and bright as some people's Twitter posts of their successes might make it seem to be. Those are just the good times and then there are the bad times. Normally, PhD students do not talk about their struggles. And I think that could be really deceiving for a young student. I had a bad time during my academic career, and it was followed by a good time.

When I started my PhD at Iowa State, I rotated through three labs. This was back in 2011, and there were many labs with money, but at the same time, there were many without money. Ultimately, I chose a lab, a neuroscience lab, that had money. It was not my first choice. I had heard bad things about the lab. At that time, I did not think I had an option.

Advisor: The PI of the first lab was getting away with a lot of things. As soon as I started, the PI would start off her mornings yelling and screaming at her students. She was condescending and then I learned that she never graduated any students. She was extremely manipulative, and there are so many distressing stories about her. Unfortunately, nothing happened to her because she was the wife of the chair of the department.

Environment: This exposed me to the political aspect of academics. I did not know academics could be like this. There were groups who supported this person and groups who did not. This PI was actually investigated during my time in the lab because she had a student who had an accident related to emotional health and wellness. All of the PhD students in her lab, including me, were called and interviewed, and in the end nothing happened.

A fresh start: After two years with this advisor, I decided the program under this PI was bad for my mental health. At that time, I was diagnosed with depression, and so were my lab mates. I had had enough. I spoke to a PI, who was in the same department. I knew this PI did not have any money. I went into his office and I told him I wanted to switch into his lab. He said that he did not have any money. I told him I would take on a teaching assistant role and make my own money. I said to him, "I need you to rescue me from this situation." My new PI agreed.

This new lab, despite the fact that I was not funded, was full of kindness and patience. It was so nice to be out of the mental shackles and not have to walk on eggshells. To support myself financially, I was driving to Des Moines, Iowa to teach. I actually really loved teaching. From Iowa State to Des Moines is approximately 35 minutes each way. I was committed to making it work. It was not easy to fund myself but it was worth it.

New advisor: My next advisor was very kind and patient, and I trusted him. I was so grateful for his accepting me into his lab.

Project: Most of my research did not produce amazing results, but I still got a number of publications. I collaborated and got a bunch of coauthorships, and my PI sent me to different conferences.

Career steps: In 2019, I finally defended, which capped off my seven year PhD. At that time, I wanted a teaching position, but I learned that getting Visa sponsorship for these positions is more difficult. I decided that I would get a postdoc.

My PhD advisor recommended a postdoctoral position at Tufts University and I worked there for one and half years. I really enjoyed the position, but the PI ran out of funding and needed to leave.

Next career move: When I applied to new places I asked my previous PIs at Tuft's and Iowa State what they thought of the lab. I trusted them so much. For example, I had an offer at Harvard with a PI. My previous PIs spoke to this new PI professor. He told me that he did not recommend this program for me and instead recommended another PI at Cedars-Sinai Medical Center. Despite the fact that she is a new PI and she is setting up her lab, I decided to listen to his advice. I took the postdoc and have been very happy.

I moved to LA in 2021 during the height of the pandemic. It crossed my mind to start research in COVID-19 but being an international person I am very limited in what I can do. I need Visa sponsorship and I am very restricted. In total, I have been at Cedars Sinai since the start of 2021 and it is going pretty well. I am going to a lot of conferences, and I am getting papers published. I like the PI, environment, and research.

Future career: In a year's time, I will start to look at my next career step and I will not be looking for tenure track faculty positions. I have decided that I do not want to be an academic. I do not like the vast amount of pressure on the PIs to not only set up their lab, but also to write grants, manage students and projects, design experiments, and publish papers. Instead, I would like to be in a staff scientist position in an academic or scientifically rigorous institution.

All that said, I am not geographically limited to the United States. I am even interested in going back to India. I have not ruled that out. My family is supportive of whatever I decide. I feel it is important that people know about the life of a PhD student. I am thankful that it is talked about on Twitter.

Advice for future PhDs: I would tell a new PhD to not compare yourself with others. In most cases, having things not work out is part of the process. I would say do not get discouraged and in the end focus on making your situation better. In addition, I would say having a PI who is supportive of you is more important than anything else.

Key Insights:

- For a successful PhD, it is more important to be in a supportive and healthy environment during your PhD than to only be in a lab that might have funding.
- Mental health is above all more important than any other part of your PhD and it is not uncommon to seek help from a counselor. In fact, most PhDs seek out professionals to talk with during their PhD to gain perspective on their situation.
- Visa sponsorship can limit career choices, but if approached proactively, there can be more opportunities.

Transition Story: Jesminara Khatun, PhD

PhD: Indian Institute of Science (IISc)
First position out of PhD: postdoc at Cleveland Clinic Lerner Research Institute
Current career position: postdoc at Mount Sinai

Motivation for pursuing a PhD: When conducting my Master's at Calcutta University, I had a great experience and all of my peers were considering doing PhDs. For me, it was a natural progression to get a PhD. At that time I was considering whether I wanted to go to the United States for a PhD or stay in India. I was thinking that I might be able to stay in India, but my ultimate goal was to do academic research in the United States. I was not sure whether going for my PhD or going after my PhD was more advantageous.

When I applied for my PhD, I had to conduct two rounds of live interview tests. The first test was knowledge-based, and then after passing that, I took the second live test, which was problem-solving focused. The exams were quite difficult and many students did not pass them all.

I was able to pass them and I got an offer to go to a premier institution, the Indian Institute of Science (IISc). Even at this time, I considered waiting to apply to the United States for PhDs, but I decided to go take this opportunity because it was a good one.

After my interviews and acceptance, I listed my top PIs and my PI listed his top students. I did not get my top choice of PIs. This would be the beginning of some trouble for me during my PhD.

Advisor: My PI had been a PI for about 14 to 15 years and I heard that he was not that focused as he had been at the start of his faculty appointment. Right away it was obvious that the advisor did not have the drive that got him the original position and reputation that he had.

How to Make Your PhD Work: A Guide for Creating a Career in Science and Engineering,
First Edition. Thomas R. Coughlin.

Environment: Despite being new to my PhD program, I immediately noticed some warning signs. There were some PhDs in the program doing their PhD for about 10 years. Most other people in the lab were frustrated with their work and were not collegial to one another. In fact, in many instances people were disrespectful and people would even get into yelling matches with one another. It was extremely toxic. Despite people's seniority in the group, it was difficult to order small consumables without consulting my boss. It was very tightly controlled.

Project: When I started my PhD, I wanted to do eukaryotic or metabolic transcriptional research. Despite my desires, my PI worked on a model strain of yeast that was only used by a small number of people, and when I researched that strain of yeast, there were only 20 other papers on this work. In some ways, it was not bad because the yeast can use methanol as a carbon source and has an enzyme that is highly produced and tightly regulated. Because of these unique combinations of factors, researchers use the yeast to express other genes. It was not totally uninteresting but it was not at all contemporary. I remember thinking that spending my PhD doing this type of research would be really challenging. There were people doing more interesting work all over the world. I wanted to be on the cutting edge of science.

Toxic culture: During my PhD, I did like the quality and culture of the group. With people shouting at each other, and demeaning one another, I decided to go to my PI to raise concerns about this to him. It was not a great situation. When I told my PI about these things, my PI did not take it in a good way. He thought I was speaking out of turn. There was a hierarchy in place and this was not received well. I was not supposed to speak out about how things were. It was very oppressive.

During the beginning of my PhD, there was a senior in the lab that helped me and I was able to understand the projects. He helped me pick up the main techniques. After he left in my second year, I felt very alone and did not feel like I had someone to discuss the work with.

Despite others in my PI's lab taking 10 years to finish, my PI had me finish in five years and four months. I was glad to get out. I did not like that environment.

Publications: In total, I published one paper as a shared coauthorship with two other first authors. That's the only paper I had from my PhD. I did other work and submitted it to my PI in a publication-ready form when I graduated, but it has not been published since.

Career support: There were a few career fairs, but besides that, there was no support at my PhD institution. Instead, most people went into postdocs after their PhD.

For me, I was motivated to get out of India and pursue research in the United States. I looked for postdocs at well-known universities and found PIs to apply to.

I found a postdoc opening at Cleveland Clinic Lerner Research Institute for my first postdoc. I liked that research the most in my career thus far. I was there for one and a half years and then shifted to Rutgers University in New Jersey because my husband just moved there for a job from India as well. It has been a year at Rutgers and it has not met my expectations for growing me as a scientist and I am going to be moving to Mount Sinai in New York, NY to do mitochondrial research.

Ultimate goal: At the start of my graduate experience, my main goal was to do research in the United States. At present, and at this point in my career, my main goal is to do research I am excited about in a great environment. I do not see myself working as a PI. Instead, I would like a different kind of position. I am interested in pursuing scientist positions in pharmaceuticals.

Advice for future PhDs: I recently had a conversation with my cousin who is a Master's student in India and wanting to do a PhD. Everything in a PhD is different from all of your training. In a PhD you are not just given a job and told what to do. We get paid way less than other people and work more autonomously without clearly defined job descriptions. I wish I knew that. Being in academia, it is very difficult to get paid well.

Key Insights:

- Toxic cultures during PhDs can be difficult to overcome. Setting clear boundaries and defining for yourself what you are comfortable with can help you navigate these challenging situations. For example, learning the benchmarks that you need to graduate can help you find out if you will be able to graduate or if it is better to leave early.
- When in academics, it is important to ensure you are in the best environment possible so that you can achieve your goals.

Part II

Your Academic Path

5

Overcoming Academic Obstacles

5.1 Proactive Steps

In the last chapter, you discovered how your PhD is going and you got an objective perspective of your PhD by analyzing your advisor, environment, and project, using the diagnostic test. The diagnostic test is a tool to objectively assess your PhD and identify its strengths and weaknesses. If your score in the last chapter was "great" or "moderate" in all three categories, then you are in a competitive PhD and in a good position to pursue an academic career. By contrast, if you scored a "poor" or "terrible" score in your diagnostics test, you might be in a disadvantaged position to apply to academic career positions. However, in this chapter, we are going to review strategies to improve your experience in the three categories (advisor, environment, and project) to help you improve your circumstances and become a proactive PhD.

The strategies in this chapter will provide you with methods to help you navigate your PhD and overcome challenging situations.

5.2 Personal Leadership

If you are reading this, you are either pursuing a PhD or working in a postdoctoral position, which means you have already taken charge of your own life and responsibilities. Personal discipline and leadership likely come naturally to you. What are some characteristics of personal leadership?

- Taking initiative,
- Defining personal goals or a mission statement, and
- Having courage to stand up for yourself in the face of challenges.

How to Make Your PhD Work: A Guide for Creating a Career in Science and Engineering, First Edition. Thomas R. Coughlin.
© 2024 John Wiley & Sons, Inc. Published 2024 by John Wiley & Sons, Inc.

5.3 The Importance of Your Advisor

In all the PhD interviews that we have conducted, the biggest impact on a PhD's experience was always the PhD advisor. In all of these cases that we have discussed, the PhD advisor did not only impact their progress, but they also impacted the environment and their project. The PhD advisor had a direct impact on what a PhD could achieve.

5.3.1 Importance of a Good Relationship

In the real PhD transition stories in this book, the PhDs describe the importance of the relationship with their advisor. PhDs who describe enjoyable experiences had great experiences with their advisor. These PhDs did not talk about challenging situations and a lack of support. A healthy dynamic with your advisor will shape the entire PhD or postdoc experience.

Recognizing when you have a great relationship with your advisor, a relationship that needs work, or one that you should walk away from is critical to getting the most from your PhD experience.

5.3.2 Working With Aloof Advisors

Aloof or absent advisors tend to neglect the work that you do or are not there for guidance. In these cases, sometimes the best solution is to move on from these advisors. In most scenarios, the faster you are able to identify that your advisor does not care about your development as a PhD and is not there for you, the faster you are able to find someone else who can train you. The purpose of the PhD is not to teach yourself everything. The goal is to learn and grow into a critical thinker in the field. This cannot be accomplished with an advisor who is not there. We are not talking about the advisor who is too famous or well known in the field that they do not have time. No, those advisors usually have a system in place (e.g. a network of postdocs and lab technicians) to provide support for you. Instead, I mean the advisors whose attentions are elsewhere and who are not there for you to do their job. It is rare to end up in one of these scenarios, but if you are in one of these, be aware! Working in an aloof advisor's lab can be a spiraling waste of your time.

5.3.3 Advisors Under Pressure Can Be Difficult

In some cases, your advisor might be working very hard to obtain funding or be under pressure to publish in order to receive tenure. The scenarios might make your advisor try to micromanage you or check in on your work too frequently.

Micromanagers can lead to a loss of creativity and increased stress. In these cases, take time for yourself to go on a lunch break or coffee break. Going for a walk and getting away from these tense situations can remind you of the levity of life, thereby increasing your creativity. Overall, just take time for yourself. Over time, the advisor should gain trust in your work if you stay consistent. Through consistency, they will begin to feel that they can trust you, and you will become a partner in their work. Just try to be a consistent worker, and if that does not work, you may want to eventually ask them for some time to work with music so that you can focus.

Managing up is one important skill to develop during your PhD that will help you with the rest of your career. No matter if you work in nonacademics or academics, managing up will help you understand how to build a strong rapport with your manager. In addition, managing up will also help your manager to see your progress.

When understanding what managing up is, it is important to understand what managing up is not. Managing up is not:

- Supervising or giving your advisor commands,
- Speaking over your advisor or going over their heads,
- Evaluating or judging your advisor's leadership style,
- Changing your advisor into a better leader, and
- Challenging your advisor about their decisions.

Instead, managing up is:

- Managing the relationship you have with your boss,
- Letting your manager know that you are on their team,
- Demonstrating that you want your advisor to succeed because that also means you will succeed,
- Learning your boss's management, leadership, and communication style,
- Increasing awareness of your own work and communication styles, and
- Adapting and aligning your work styles to form a productive relationship focused on mutual growth and understanding, work productivity, and development (Figure 5.1).

5.3.4 Managing Up With Your Advisor

Managing up refers to building a positive working relationship with your supervisor or manager and effectively communicating with them to ensure that you both are aligned on priorities and goals. It is one important skill to develop during your PhD that will help you with the rest of your career. No matter if you work in nonacademics or academics, managing up will help you understand how to build a

Figure 5.1 By managing up with your own PhD experience, you can earn respect from your advisor while making progress with your own work.

strong rapport with your manager. In addition, managing up will also help your manager see your progress.

Managing up means having strong relationships with key superiors, like your advisor (Abbajay 2018). It is important to take the time to understand what the goals of your advisor are, as knowing these will help you excel in your job and demonstrate how you can be a supportive team member. Regularly communicating with your advisor helps ensure that everyone is on the same page and can go a long way toward strengthening relationships. As well as bringing problems to their attention, it is also beneficial to provide potential solutions whenever you can. Being proactive about finding ways forward rather than just pointing out issues is welcome by most supervisors, as it shows that you are taking initiative in trying to resolve things efficiently. Furthermore, being a collaborative team player instead of an individual contributor encourages cooperation and trust, while keeping your supervisor informed will always be appreciated. Ultimately, successfully managing up is an essential part of working life that pays dividends when it is done properly.

Overall, managing up effectively requires clear communication, proactive engagement, and a willingness to work collaboratively toward shared goals. By building a positive working relationship with your supervisor and effectively communicating with them, you can help ensure that you are aligned on priorities and working toward shared objectives.

5.3.5 Improving the Relationship With Your Advisor

The quote, "Don't ask for respect, earn it," applies here but I tend to like the other version of this mantra. Instead of earning it, "don't ask for respect, take it." As a PhD student, you might have to make it quite obvious that you are contributing to the lab or group and the advisors' vision. Try to understand your advisor's vision and make your relationship with them better. Of course, there are always the advisors that will not make time for you. However, if you show your own personal leadership and define benchmarks for what you will tolerate for yourself, then you might be able to make your relationship with your advisor better, thus improving your PhD better.

There are a number of ways to improve the relationship with your advisor and start to work with them in order to achieve respect and gain momentum with your PhD (Figure 5.2). For example, getting one on one time with your advisor might not be on their radar. However, if you make it a priority to help them get to know you with a leisurely walk or a cup of coffee, you might be able to build rapport which helps you work more easily with them. Another way to understand your advisor's job, and learn about the career as an academic is to try and understand your advisor's plans and goals for bringing in grant money. In this way, by seeing their goals, you can more pointedly ask what they would like help with, or more importantly, ask for a specific project you would like to work on. By discussing their grant goals, you can build rapport with your advisor and, simultaneously, help you gain writing skills in the process. In addition to helping with grants, offering to write a review paper is a great way to make progress with your PhD and earn favor with your advisor. A review paper helps you get up to date on the field you are working in while also giving you a potential publication in the field you are working in. And your advisor will appreciate the initiative while also gaining themself a new publication. Overall, if your advisor's research plans are not laid out for years to come, having a conversation with them about research ideas will help you see how your advisor thinks. You might even be able to have some scientific credibility and influence where the research goes.

5.4 Navigating Challenging Environments

Navigating challenging work environments can often feel overwhelming, but there are proactive steps you can take to manage the situation and maintain your productivity and well-being. Identifying the source of the challenge is a great start. Once you know what it is that is causing difficulty, you may have clarity on how to address it.

Going for a coffee
Informal conversations over coffee can help build a good rapport with your advisor.

Discussing grant goals
Asking to discuss grant goals for the year shows your interest in their research and commitment to your own project.

Offering to write a paragraph or conduct a literature review

Offering to write a review paper
Offering to write a review paper can showcase your writing skills and expertise in your field.

Discussing future research ideas
Discussing future research ideas and offering your own can demonstrate your enthusiasm for the field and potential for collaboration.

Figure 5.2 To work on the relationship with your advisor and become better at your own research and writing, you can work on your relationship with them in a number of ways. Going for a coffee, discussing their grant goals, offering to write part of a grant, writing a review paper, or discussing future ideas are a few ways to help foster a positive, working relationship with your advisor.

When possible, focus on what you can control, such as improving your understanding of projects and setting goals for yourself. Additionally, building relationships with positive colleagues can be an effective way to foster support and develop potential solutions. Practicing self-care during this time is also important, taking breaks throughout your day, setting boundaries with challenging supervisors, and trusting in your professional abilities can help bring renewed energy back into the workplace. Finally, locating the counseling center or graduate career support center at your institution is a great step when seeking advice or guidance.

Taken together, navigating a challenging work environment requires a combination of self-awareness, communication skills, and self-care practices. By taking steps to manage the situation and prioritize your well-being, you can stay productive and maintain a positive attitude.

5.4.1 Overcoming Toxic Environments

Toxic environments need to be avoided. If your environment is toxic, the key to your success is going to be getting away from the bad environment to obtain perspective. If you are in a lab with an aloof advisor, and your lab mates are sabotaging your experiments or using all the supplies, thereby hurting your chances of success, then you need to carve out time when you can do your experiments and get them done without interruption or negativity. It is very hard to think creatively in a toxic or stressful environment, and therefore, you need to find time for yourself.

There are multiple strategies for overcoming toxic environments. You may try getting to the lab two hours early or going to the lab in the evening after the toxic lab members leave. In many labs, going in early or after nine to five work hours is completely normal and acceptable. Most of the time, the advisors are not aware of the general toxicity of the lab, and it is important to take control of the situation you are in. Advisors care more about you getting your work done and are not generally concerned about when you do the work.

Most university PhD advisors are not going to turn you down if you explain your situation about the lab being busy or chaotic during the day and that you think your experiments may be easier to get done during the evenings. As long as you make it clear that it will help you do better work then you should be good to get approved for working nights. In the end, the advisor's primary concern is getting the work done so they can present it to the grant's committee.

5.4.2 A Lack of Funding

If you are conducting a PhD that needs research funding to pay for the work, then a lack of funding can be detrimental to your work. In these cases, your advisor is likely working on applying for funding. If they are not desperately working on funding grants then, then you may consider tossing in the hat and finding a new advisor because getting funding is their job. In PhDs, where your advisor is applying for funding, the best thing to do here is to ask your advisor if you can apply for some PhD or postdoc level grants. In these cases, you can let them know that you are looking at governmental grants and are considering obtaining funding for some fellowship program. Alternatively, a first endeavor may be to ask the advisor if you might be able to help with their grant proposal.

When working for a new advisor it is very often the case that the advisor is still not well known in their field. In order to establish themselves in their research field while they are still inchoate, they might find it useful to write a review paper. Review papers can be an excellent way for PhDs to make good use of their time during funding droughts.

Review papers offer a win–win scenario for PhDs. In writing a review article, a PhD gains many advantages. First, you can obtain an in-depth overview of the field, oftentimes becoming an expert in the field and where it is going. Secondly, a PhD can garner experience in writing. Third, you can get a publication to add to your CV and talk about to your thesis review committee. Finally, a review article is a great way to show your productivity and build trust with your advisor.

5.5 Overcoming Challenging or Uninteresting Projects

Completing challenging or boring projects can pose a significant obstacle to anyone, especially PhDs. Using the right strategies you will be able to stay motivated and productive. Breaking up large projects into smaller tasks can make them seem more manageable while setting specific goals gives you measurable progress to measure. Finding ways to make the work more interesting can be beneficial, you could listen to music, set time limits for tasks, or experiment with different methods of execution. Taking regular breaks is essential as it allows your mind to reset and refocus. Additionally, speaking to your colleagues if stuck on something can provide clarity and fresh ideas. Finally, rewarding yourself after completing each task is a great way to inspire you to continue working and increase productivity overall.

5.5.1 Stagnant or Insipid Projects

Okay, you have been assigned a project that your advisor just needs you to finish. There is a lot of labor to this task and you do not know if you will get through it and it is just downright depressing.

I remember a scenario like this in my PhD. The newest PhD student in our lab was given a project that needed to be finished for an old R01 grant, and the PhD who conducted the original research had graduated. The new PhD student was originally excited for the project, but as time wore on, that PhD student grew tired of the work, which involved imaging slides of bone tissue at high magnification. Oftentimes, each slide would require 100 images and there were over 100 samples to image each with two slides. The task was daunting.

The project required him to sit at a microscope for an entire year for approximately 40 hours per week, conducting microscopy in the dark. I watched as my colleague was drained from these experiences. Amazingly, as one of the most persistent people I have ever met, he overcame this experience by focusing on what he could control. He left each day at five o'clock and was very punctual when he started. He also took breaks and walks and was very active after work with sports. He told us about what he was going through and as lab colleagues, we would stop by amidst the imaging and cheer him on when he would be at it. He also broke up the large project into small tasks and set specific goals to measure his progress. He listened to music, set time limits, and even streamed TV shows. Finally, he took regular breaks to reset and refocus and spoke to me and his other colleagues to get fresh ideas and looked to us for support.

That said, he got through it, and he seemed to look back on the experience as something he would not want to do again but he forever learned to maintain that type of balance in his life. In the process of doing this laborious research project, he showed our advisor that he was up for the challenge for difficult projects and our advisor trusted him with some new research that he was interested in. Albeit, the work was not the most pleasant experience, but my peer finished the project and got through it with a publication.

If you are working on an old project, like my peer, you may want to ensure that you are able to get on the publication as first author and you can write the manuscript. Although the work might not contribute to your research objectives that encompass your PhD, you will at least be able to have a publication to talk about with your review committee and gain experience in writing in the process.

5.5.2 Tools to Overcoming Uninteresting Projects

Another strategy to overcoming uninteresting projects is to try and come up with some research projects on your own. You might want to try and think of a few different ideas or avenues and ask your advisor if you might be able to apply for a grant to work on these projects. At the very least, this will start a conversation with your advisor about new research directions and in the best-case scenario you might get one of the grants and forge your own research path.

5.6 Conclusion

Overall, there are many ways to be proactive during your PhD. You might be able to work on new ideas, apply for fellowships or obtain training in a new area. In addition, you can use some downtime during your PhD to look into future career directions by taking courses in teaching, research, or industry careers. Your time is your most valuable asset and it is important to make the most of it.

Chapter 5 Key Takeaways

- ❖ Focus on what you can do during your PhD if you are in a negative PhD program.
- ❖ Try to work around toxic environments to get space and time away from them.
- ❖ Make efforts to gain your advisor's trust by being consistent and trying to manage up in some scenarios.
- ❖ Overcome uninteresting or recycled projects by showing interest in new areas with your advisor and applying for fellowships or grants.
- ❖ In addition, make sure you are able to get onto a publication if you are working on an old research project in your lab.
- ❖ During down times during your PhD, you can offer to write a review article, which will build trust with your advisor, while developing your writing skills and garner you a publication.

Reference

Abbajay, M. (2018). *Managing Up: How to Move Up, Win at Work, and Succeed with Any Type of Boss*. Wiley.

6

A Purposeful Postdoc

6.1 The Transitional Postdoctoral Fellowship

After graduating from my PhD, I chose to pursue a postdoctoral fellowship. In 2015, I had a limited understanding about the opportunities available to me outside of academia and assumed I could learn what was out there during my postdoc while still pursuing an academic career. As a first-year postdoc, my base salary was set by the National Institutes of Health (NIH) at approximately $45,000 and went up to $47,500 in my second year in 2016. As you can imagine, in New York City, this was not enough to live on, let alone save money for retirement, or even emergencies. However, I was not alone in this. Many PhDs earn paltry salaries in postdoc positions across the United States after graduation. And before resources like this book, PhDs had to fend for themselves to discover their own careers.

The truth is, the postdoc period is transitional, and if you know what you want and that you can get there without a postdoc, you should avoid a postdoc at all costs. Why? Because the transition into a different career outside of academics should be made before the postdoc starts. And because postdocs can be time-consuming, requiring you to spend all your time doing research and not enough time on your own career transition. And did I mention that you make very little money.

The average time spent in postdoc positions has swelled from one to six to ten years (Hankel 2022). In addition, and quite sadly, many universities label these postdocs as "nonfaculty staff" to hide the number of postdocs they have. In many cases, the postdoc is an extremely well-educated, experienced, expert-level labor resource for an incredibly cheap price. The problem of postdocs stems mainly from the root of problems within academics. It is also a result of underfunding professional labs in academic institutions.

How to Make Your PhD Work: A Guide for Creating a Career in Science and Engineering,
First Edition. Thomas R. Coughlin.

6.2 Postdoc Unions and National Associations

Despite the continued low wages and benefits for postdocs, they have started to work together. Many universities, including Rutgers University, Columbia University, and others have had postdocs who formed unions to improve working conditions and benefits. In addition, the National Postdoctoral Association (NPA) was formed in 2003 and aims to enhance the quality of the postdoctoral experience for all participants (National Postdoctoral Association 2023).

Taken together, and simply put, the postdoc is not an easy time. The postdoc position is a means to end, but it is not the end. Approach a postdoc with specific goals in mind, but only pursue one for specific reasons.

6.3 Know Before You Go

It is important to not waste any time in a postdoc. Below I have included a table of some of the right and wrong reasons to conduct a postdoc (Table 6.1). There are many good reasons to conduct a postdoc after your PhD. For example, the postdoc is an incredible time to carve out your path in academics, apply for grants like the R00/K99, and apply for academic jobs. It is a great time to give academics one last try, or to ensure you are competitive for an academic position.

I have heard many PhDs ask, "Do I need to gain more skills in a postdoc to get a job in industry?" The answer is no. You do not need to conduct a postdoc to get access to a job in industry. Postdoc training does not train you to become an industry professional. There are outside courses you can take and books you can read, like this one. But a postdoc does not inherently train you to get a job. Yes, it gives

Table 6.1 The right and wrong reasons to pursue a postdoc.

Right Reasons for a Postdoc	Wrong Reasons for a Postdoc
Bridging from PhD to an academic job by: • Honing and better defining your research niche, • Leveraging your PhD skills to conduct high-level research without the barriers that you had during your PhD, and • Writing and applying for a K99/R00 postdoctoral training grant. Moving from PhD to nonacademic job by: • Obtaining the training in a program that has proven to lead to a specific career.	Utilizing a postdoc for: • Gaining research skills for a future vague position, • Deciding what you want to do with your career, • Determining if academics is right for you, and • Learning what you like in a future career.

you more skills, but it does not train you to get a job. That, inherently comes from knowing the goal you want and going on the right path to get it.

If you fall short of academics, and you decide to move on from that path, the postdoc can be a great time to nurse your wounds from academics and move on with your career. You can use the time as a period to explore careers and transition into the job.

6.4 Obtaining a Postdoc

The search for finding a postdoc position can be long, but with the right knowledge and resources, it can be manageable. Taking a look at LinkedIn, Indeed, and other job board websites is key when attempting to find an open postdoc position. It is also worth checking with advisors, professors in your department, and even at conferences, as they may have contacts or other knowledge that could help you secure the perfect position. Taking advantage of these resources will give you access to the best job openings available, including both formally posted and ones available by word-of-mouth.

Most importantly, when searching for positions, consider your goals and find out if they align with the lab or group's goals that you are interested in.

6.5 Postdoc Interviews: What To Expect

When going for a postdoc interview, you should keep in mind that you are taking the initiative to assess whether or not the position is right for you. Do not forget that it is just as important to understand the principal investigators (PI's) expectations and communicate your goals clearly as it is for them to understand how you fit. Ask questions about what past postdocs have done after leaving the lab and find out where the lab funding comes from and how the postdoc position is funded. Additionally, make sure you inquire about being able to take your project with you once your postdoc period has ended. Also, find out how authorship is handled in the lab and get clarification on what the institution's policy is when it comes to postdoctoral positions. Here are key questions to ask in your interview:What have past postdocs done after leaving the lab?

- Where does the lab funding come from? How is the postdoc position funded?
- Will I be able to take my project with me once the postdoc ends?
- How does the lab handle authorship?
- What is the institution's postdoc policy?
- What do you expect of your postdocs?

6.5.1 Postdoc Interview Process

The purpose of the interview is for both you and the PI to determine if your skills, experience, and personality are a good fit for the position. Postdoc interviews process can include:

- 1: 1 with PI via phone or video conference,
- On-campus interview 1–2 days,
- In-person interview with PI,
- Meeting(s) with members of lab or department,
- Touring the campus, and
- Presenting your research.

6.5.2 Postdoc Interview Questions

Preparing for your postdoc position interview questions is essential for any successful application. A good preparation involves researching, analyzing, and rehearsing any possible answers to cover every area that can potentially come up during the interview. Familiarizing yourself with the research projects of the lab you are applying to, as well as understanding its specific culture, objectives, and goals will provide valuable insight into how you should answer various interview questions. By taking notes on these points and incorporating them into your mock conversations and responses, you will be better positioned to convince the interviewer of your suitability for the postdoc position.

Broadly, your interviewers want to know about your:

- Teamwork,
- Strengths and weaknesses/development areas,
- How you deal with failure, and
- Work style.

Some of the common questions are:

- Tell me about yourself.
- What do you research?
- What are the main findings of your PhD?
- What areas of our research interest you and why?
- How will your research fit in our department?
- How would you seek funding for your work?
- How will this postdoc help you reach your professional goals?
- Do you have experience mentoring graduate students?

6.5.3 Postdoc Talk

The talk during your postdoc interview is there to help the group you are interviewing with understand your research, hear how you present, get to know you in a professional way, and look for potential areas of collaboration. The talk should be approximately 20-40 minutes depending on how much time you are given. While giving this presentation be sure to ask who the audience is so you can cater the technical depth of the work to the attendees. You might not want to jump right into a deep neuroscience talk to a group of marine biologists without giving background and rationale for your work.

6.5.4 Thank You Follow-Ups

When it comes to follow-up after an interview for a postdoc position, timing is essential. You should send a thank you within 24 hours to show that you are interested and appreciate their time. In this follow-up, reiterate your enthusiasm for the position and demonstrate your commitment to the role. After the email is sent, consider sending a short and professional thank you note as well to the other people you met during your interview. Doing so will ensure that you have made as strong of an impression as possible!

6.6 Conclusion

If you are considering a postdoc, it is important to evaluate the situation and make sure pursuing one makes sense in the context of your career goals. Do not just take a postdoc because it seems preferable to the labor market that still exists. Think carefully about what you want to get out of one and whether it will be worth your time and money. The results of a good postdoc can, however, bring great rewards. The real value of taking a postdoc, as opposed to other options available to PhD graduates on the job market, is having an opportunity to collaborate with experts in your field and being able to hone knowledge and skills that cannot be acquired in any other way. So if you examine opportunities critically and thoughtfully choose a position that fits with your long-term goals, then there is no reason why they will not lead you toward success!

Chapter 6 Key Takeaways

❖ The postdoc position is a means to an end and there are key reasons to do a postdoc and not to do one.

❖ Defining clear reasons for conducting a postdoc can help you stay focused and motivated to decide when to move on or when to keep going.

❖ It is important to not only attend interviews for postdocs but also to interview the lab or group members and the advisor of that position. It should be obvious if an advisor's goals are aligned with yours. This is key to your success.

References

Hankel, I. (2022). *The Power of a PhD: How Anyone Can Use Their PhD to Get Hired in Industry*. Morgan James Publishing.

National Postdoctoral Association. (2023). About. National Postdoctoral Association. www.nationalpostdoc.org (accessed 27 March 2023).

7

Creating an Academic Plan

7.1 My Academic Experience

My undergraduate institution was a teaching college. The professor-to-student ratio was eight to one, giving me ample opportunity to learn and engage with my professors. I was not exposed to PhDs, Master's students, or staff scientists until I got to my PhD program. But what drove me to get a PhD was a yearning to learn more than what I learned in undergraduate school. I remember being disappointed in my junior year with how much I had learned. I thought this cannot be it. That summer between my junior and senior year, I worked at an internship in Integra Life Sciences at their headquarters in Plainsboro, New Jersey. It was a competitive internship and when I got there, I thought I was going to be able to contribute ideas or be on the cutting edge. Instead, the work during the internship primarily focused on singular one-off jobs in the clinical affairs department where they enrolled patients in clinical trials and created documents for patients to sign. I later learned that in order to be respected for my research and development ideas or do more than a singular job, a PhD was needed. A PhD would give me the chance to be on the cutting edge. This was academics. I had found my path.

7.2 Energy Does Not Always Equal Results

During my PhD, I learned that academic institutions offer different types of professor roles. There are research-focused professor positions and other professor jobs that are teaching-focused. Then, there are balances between them. In this chapter, I will review the different types of jobs while explaining the application and interview process.

How to Make Your PhD Work: A Guide for Creating a Career in Science and Engineering, First Edition. Thomas R. Coughlin.
© 2024 John Wiley & Sons, Inc. Published 2024 by John Wiley & Sons, Inc.

Table 7.1 PhDs can be hired in faculty positions in doctoral universities, Master's colleges and universities, and baccalaureate colleges. These three types of institutions have more specific classifications that separate them from one another and describe the level of research, the size of the university, and its level of focus on teaching.

Doctoral Universities	Master's Colleges and Universities	Baccalaureate Colleges
• R1: Doctoral Universities – very high research activity • R2: Doctoral Universities – high research activity • D/PU: Doctoral/Professional Universities	• M1: Master's Colleges and Universities – larger programs • M2: Master's Colleges and Universities – medium programs • M3: Master's Colleges and Universities – smaller programs	• Arts & Sciences Focus – teaching focused • Diverse fields – teaching focused

7.3 Types of Institutions That Hire Academic Faculty Positions

There are three types of institutions that hire faculty positions, including doctoral universities, Master's colleges and universities, and baccalaureate colleges. Within each of these types of institutions, there are classifications of institutions, described in Table 7.1 (Bloch 2010).

7.4 Types of Academic Faculty Positions

Within each of these three types of institutions there are different types of faculty positions, including (Bloch 2010 and Office of Career & Professional Development 2012):

- **Teaching-Intensive Careers in Academia:** A primarily teaching faculty position in a research university, liberal arts college, or community college.
- **Combined Research and Teaching Careers:** Faculty at a liberal arts college or university whose job includes both research and major teaching responsibilities.
- **Principal Investigator in a Research-Intensive Institution:** Independent researcher at a medical school, private research institute, and government lab or university with minimal teaching responsibilities.
- **Research Staff in a Research-Intensive Institution:** Staff scientist or researcher in academia or government, lab manager, director of a multiuser research facility in an academic institution.

7.5 Professor Promotional Ladder

You have likely heard of tenure track faculty, and you might have seen differences in professors' titles. There are specific career ladders for academics, where PhDs progress from:

- Assistant professor, to
- Associate professor, to
- Professor.

7.5.1 Assistant Professor

As an assistant professor, one not only provides teaching duties, but they are also involved in research and professional activities as part of their job. While the workload varies depending on institution and department, tenure track professors may be involved in evaluating student performance, preparing lectures, conducting lab classes and conferences, taking part in committees, researching, and publishing a variety of materials – all while progressing to fulfilling requirements for achieving tenure. Thus, assistant professors play a key role in expanding the educational knowledge of future generations as well as offering valuable mentorship opportunities.

7.5.2 Associate Professor

An associate professor is a mid-level professor occupying a tenured position between an assistant professor and full professorship. The title of an associate professor is typically given to those who have undergone a rigorous tenure process, lasting between five to seven years. During this period, the faculty must demonstrate exemplary research or teaching skills in order to become tenured. Associate professors oversee the creation and maintenance of curricula, teach classes according to their academic specialties, and engage in independent scholarly research to stay up-to-date on their respective fields. In short, associate professors are important as they provide instructional leadership, mentor students, and promote academic excellence at universities across the globe.

7.5.3 Full Professor

Being appointed as a full professor signifies the highest possible career level an academic can achieve through tenure. While achieving this rank often takes many years, it brings remarkable opportunities and benefits including increased salary and the freedom to pursue research beyond one's doctoral studies. In addition to the status that comes with being a full professor, many institutions have additional honorary titles or positions they can grant, such as distinguished professor or chair professor. Those with distinguished professorships typically receive higher salaries and greater acclaim as well as reduced teaching loads so that more of their attention is given to research. Ultimately, having achieved full professor status shows great dedication and commitment to an academic field that helps move it forward.

7.6 Tenure Track

Tenure status is given to assistant professor positions to describe the level of commitment that the academic institution has made to this position. There are nontenure track, tenure track, and research track positions.

- **Tenure Track Faculty:** engage in a diverse range of activities to remain competitive, having a high degree of excellence in at least one area, teaching, research or service, and adequacy in the remaining two.
- **Nontenure Track Faculty:** still need to showcase skills with their main mission being teaching and contributions to service with attention paid to excellence in the primary mission and adequacy in the other.
- **Research Track Faculty:** are more focused on their research initiatives, though many departments also require them to demonstrate performance through service too.

The tenure track is a crucial pathway for college professors to advance in their academic positions and gain job security. Most American universities utilize the tenure track as a way to promote faculty, which typically begins as an assistant professor and can move into an associate professor and then professor if successful.

7.6.1 The Tenure Process

Academic institutions post jobs looking for either tenure track, contractor, or term positions. In tenure track jobs, you will be required to meet a certain eligibility criteria to stay in the position over the course of a five or seven year review period (August 2022). During this five to seven year review process, a professor is often denoted as "going for tenure," or "tenure track" faculty.

Being on the tenure track is a big job and a milestone for any professor. In order to go up for tenure review, which is typically required within six years of entering their position, professors must put together a comprehensive dossier containing information about their research, service, and teaching. This includes details like their CV, list of publications, statement about why they should be granted tenure and a comprehensive teaching portfolio including awards and grants earned at the university. Additionally, an external review from 5 to 10 senior scholars in the field will also be conducted to not only get an assessment of the professor's work but also to judge how much impact it has had on the wider scholarly community. After all this evaluation has been done then the final decision will be made as to whether or not the professor receives tenure.

The process of securing a tenure track position and then accomplishing academic promotion can be complex, often requiring multiple rounds of reviews. Once all external letters of review have been collected, the departmental tenure committee will review the professor's dossier, followed by a recommendation from the department head. At that point, the tenure dossier moves up through the academic hierarchy to the faculty dean, campus committee, and eventually to the provost for a final decision. If awarded tenure, many professors are also promoted to associate professor. With further experience and sustained commitment to their field of research and teaching, associate professors are again reviewed five or seven years later for possible promotion to full professor. For those on a nontenure track or research track path, promotions can also occur along similar lines although there may be different criteria applied in the approval process.

7.7 Deciding on a Research Focused Position

What might make you want to do a research professor job? Well, a research professor takes a rare breed. It combines the rigor of teaching, research, and advising, and the professor is relied upon by the university to run a mini-research startup. That research professor balances hiring and advising PhDs, obtaining research funding, and teaching courses. If you are interested in all that and want the research obligation, then this is what you aim for. However, if your primary interest is teaching, then you would aim for a teaching professor role with very little research obligations.

When looking for these jobs, the type of school dictates the type of professor that they look for. You might know what type of school your undergraduate institution is or was. A smaller liberal arts school with smaller budgets and private funding, may rely more on tuition for revenue and employ teaching professors to cover the course loads. This is very typical of community colleges. In contrast, a research professor would work predominantly at the larger institutions, which are typically the universities with bigger budgets and are in general, more well-known. Either way, you can make a great and fulfilling living doing either.

7.8 Deciding on a Teaching Focused Position

On the other hand, a teaching professor is hired by a university to teach and is paid to cover a large number of courses or class units. This is typically 4–5 a semester. In addition, the teaching professor will also be responsible for advising students.

For these tenure track roles, the professors will also have a tenure review committee who will evaluate their tenure package or portfolio. There are tenure track teaching professor roles that still require a publication every one or two years. However, these publications can be less research-based and more review-based and are not judged as stringently for contribution to the field as journal articles published by a research professor. In addition, the teaching professor is judged by their contribution to the institution. They are evaluated on their teaching, their advising, and their general involvement in leadership activities in the university.

7.9 Comparing and Contrasting

Each institution has its own metrics and standards for achieving tenure. Although I have not tried to achieve this myself, I have a few anecdotes from my experiences and familiarity with the process. Here are three stories of tenure track faculty.

R1 University Story

Dr. Jones was conducting research in a tenure track research professor position at a medical institution. In this role, Dr. Jones' salary was paid by her research and the institution. Seventy-five percent of the salary was paid by the department that hired her at the medical institution and 25% of her salary could come from grants that Dr. Jones brought in. It is a common practice to build some salary pay into a government grant proposal. Upon employment, the medical research institution provided Dr. Jones a research grant and then gave the professor three years to get her lab up and running.

In her third year, Dr. Jones had published a few papers and had a few in the works, but the National Institutes of Health (NIH) R01 research grant proposal that she had hoped for, was not receiving scores high enough to deem them fundable. Dr. Jones had already submitted the grant, and after the first round, she was invited to submit it again. At this time, her grant was not funded.

In the coming weeks, after the NIH decided against funding her proposal, the tenure review committee turned her away and she was asked to leave. Dr. Jones fortunately had ties at another university and was invited to work there while she tried a few different approaches to her research. The postdocs and researchers she had hired were left to finish their work or move with her and the medical institution closed her lab after the final research was completed by her postdoc.

Key Insights:

- In some medical institutions, receiving funding is vital to receiving tenure.
- What do you have to lose going for a research position at a university? The only thing is to not succeed, which is much better than never trying at all.

R2 University Story

Dr. Brighton was working in a brand new bioengineering department for seven years before he was told that he did not receive tenure. His research had met all the criteria of the department, his students had graduated with PhDs, and he had published significant work in a well-enough known niche of research to be considered a significant contributor to the field.

Despite the research and efforts, Dr. Brighton was denied tenure due to reasons that are a little challenging to understand. The bioengineering department at the time was nestled inside the mechanical engineering department and thus, funding and career decisions for the department were not solely made by the bioengineering department. So, when the decision for the tenure was up for review, the mechanical engineering department decided to go in a different direction because they wanted more faculty to do more multidisciplinary work for more potential collaborations.

Key Insight:

- It is important to be aware of the department's standing in the university and try to anticipate if there may be any changes in priorities of the university in your given field of research.

Baccalaureate College

During Dr. Grom's PhD, she learned early that she did not want to pursue research and instead wanted to teach at an institution focused more on teaching. In order to make herself more competitive and to gain experience in teaching, Dr. Grom began teaching a lecture class in the last two years of her PhD. She also received a certification in teaching excellence from the university and applied to many schools for teaching professor jobs. After three on-site interviews, Dr. Grom had a two offers.

Dr. Grom was a newly hired tenure track teaching professor at a private institution in the Midwest. Dr. Grom was a mathematics PhD and was hired to teach in the mathematics department. Upon entering the institution, she taught five classes each semester and was an assistant with the honors club. After two years of work, she decided to change institutions after her alma mater opened up. Dr. Grom settled there and was able to carry over her tenure track work and progress at her new institution. In three years, upon being up for review of tenure, Dr. Grom received tenure from her review committee and continued to work teaching four to five classes each semester.

Key Insight:

- Taking steps to gain teaching experience during the PhD led Dr. Grom in the direction of teaching professor and helped her establish herself as a teaching-ready candidate in job interviews.

7.10   Becoming Competitive for R1 Positions

In today's competitive landscape, striving for a faculty position in an R1 research institution is no longer just about having a PhD. There are other metrics needed to become competitive – such as demonstrating your ability to secure external funding and presenting your work both on a local and national level. It is invaluable to take advantage of resources such as the Office of Grants and Fellowships, applying for postdoc awards, career development/transition awards, young investigator awards, and networking with like-minded people in order to demonstrate competency beyond that attained through a PhD alone.

7.11   Becoming Competitive for R1 Positions

Being competitive for a teaching-focused professor job requires showing evidence of teaching experience. This may include previous TA duties, being an instructor of record, and pursuing education professional development opportunities. Utilizing your personal institution's teaching resources is often beneficial in building this experience, as well as publishing any findings or research related to the topic. While a postdoc may not be necessary for you to complete, it can serve as an opportunity to further one's skills and knowledge and help them stand out among applicants.

7.12   Combined Research and Teaching

For any hopeful research and teaching professor, the key to becoming competitive for such a job is to not only be able to teach but also have a well-developed research history. Undergraduate students nowadays are looking for experiential learning opportunities and what better way than through the collaboration of research pursuits with their professor? Furthermore, if you can get your work published, secure funding for your projects, and show a willingness to involve undergraduates in your lab pursuits, you can certainly become an attractive candidate for this type of job. As the higher educational landscape continues to require professors that are active in their own efforts as well as in engaging students, showing potential employers evidence of this mission is paramount.

7.13   The Professor Application

There are a few key components of professor job applications, including:

- Cover letter
- CV
- List of recommendations/references
- Research statement
- Teaching statement
- Diversity statement
- Teaching portfolio

For a teaching professor position and research professor position, different weights are placed on the research statement and teaching statement. Teaching professor hiring search committees look more closely at the teaching statement and make sure it more closely aligns with the direction of their university and if it aligns with their general mantra. With respect to a research professor's job application, the research professor search committee will look more closely at the research application and plan. Let us take a look at what each component of the application consists of.

7.13.1   The Cover Letter

The cover letter is a traditional document that explains the intentions of your application (Career Services 2020). The first paragraph of your cover letter should clarify the open job and job number (if there is one) that you are applying to, where you are in your PhD, postdoc, or career, and then explain why you would be a good fit for the position. In the next paragraph, outline your strengths – particularly those that are most relevant to the position. For example, if it is a research job application, you should briefly describe your research plans, grants you have received, and your plans for your lab and how they fit with their department. In the final paragraph, provide your contact details so that they can contact you if they need more information. End the cover letter by mentioning that you look forward to speaking with them about the opportunity.

7.13.2 A Curriculum Vitae (CV)

A curriculum vitae, commonly abbreviated as CV, is a comprehensive document that provides an overview of a person's academic and professional history (Indeed Editorial Team 2023). Unlike a resume, which is typically limited to one or two pages, a CV can be several pages long and includes several different sections of information (Figure 7.1).

CVs are commonly used in academic and research settings, as well as in certain industries where extensive experience and qualifications are highly valued. The detailed information included in a CV allows employers or academic institutions to evaluate a person's full range of skills and achievements, making it an essential document for certain job applications.

With the CV being the one document to them all, the CV is your place to store all of your accomplishments. It can act as a database for you to keep track of the work you have done.

7.13.3 The List of References

As you progress through your PhD and postdoc you will accumulate a series of advisors, mentors, and collaborators that can give you a referral for your next career. This list of references typically includes members of your PhD thesis review committee, your advisor, and your postdoc advisor, if you have one.

7.13.4 The Research Statement

The research statement is typically between 3 and 5 pages long and includes your future plans for your work (Bhosale 2022). This research includes references to your previous work to explain why you are the right person to do this work. Furthermore, highlighting your previous work demonstrates why you are respected enough in the community to receive funding for this work. At the end of the statement, write your plan for applying for grants and the immediate steps you would take if employed.

Most importantly, the research statement shows the career search committee how you plan to be immediately productive at their university. It shows that you

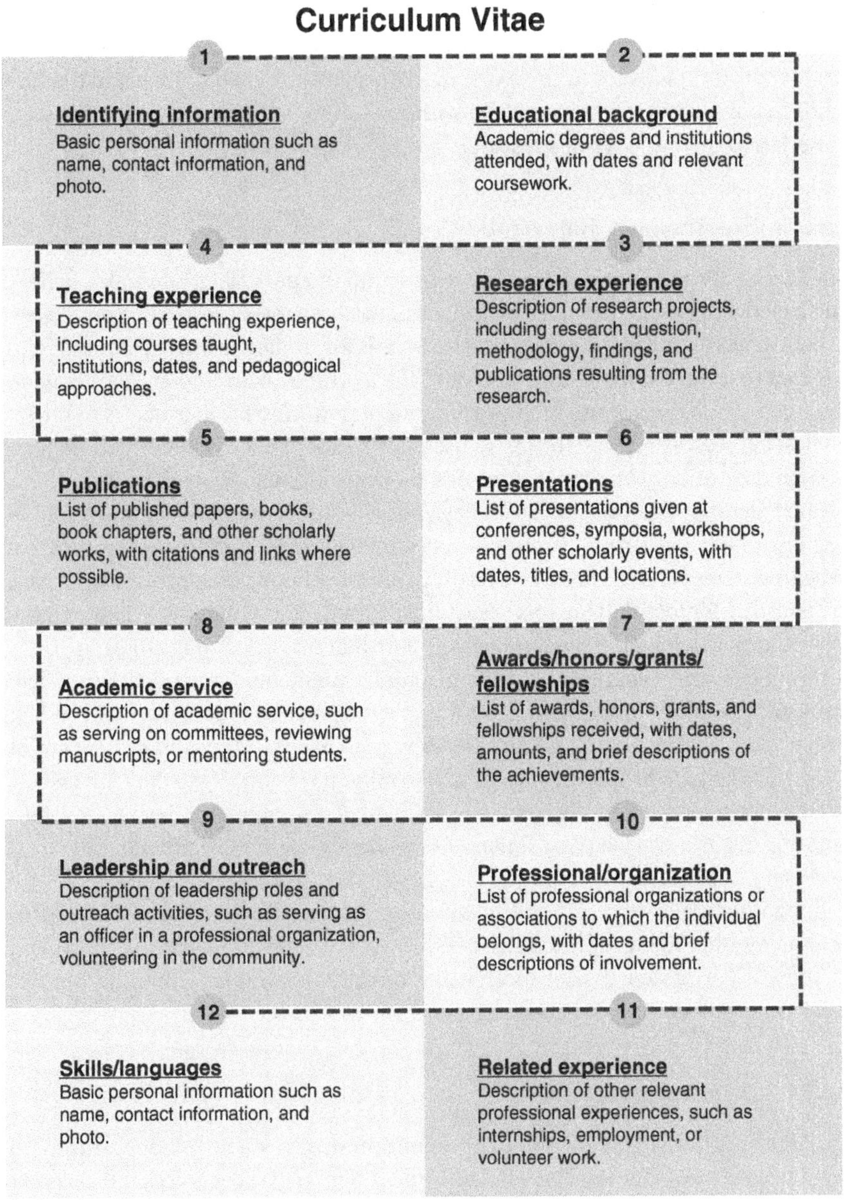

Curriculum Vitae

Identifying information
Basic personal information such as name, contact information, and photo.

Educational background
Academic degrees and institutions attended, with dates and relevant coursework.

Teaching experience
Description of teaching experience, including courses taught, institutions, dates, and pedagogical approaches.

Research experience
Description of research projects, including research question, methodology, findings, and publications resulting from the research.

Publications
List of published papers, books, book chapters, and other scholarly works, with citations and links where possible.

Presentations
List of presentations given at conferences, symposia, workshops, and other scholarly events, with dates, titles, and locations.

Academic service
Description of academic service, such as serving on committees, reviewing manuscripts, or mentoring students.

Awards/honors/grants/fellowships
List of awards, honors, grants, and fellowships received, with dates, amounts, and brief descriptions of the achievements.

Leadership and outreach
Description of leadership roles and outreach activities, such as serving as an officer in a professional organization, volunteering in the community.

Professional/organization
List of professional organizations or associations to which the individual belongs, with dates and brief descriptions of involvement.

Skills/languages
Basic personal information such as name, contact information, and photo.

Related experience
Description of other relevant professional experiences, such as internships, employment, or volunteer work.

Figure 7.1 The CV is made up of specific sections of information that describe your career in academics.

have thought about how you will obtain funding, who you might collaborate with, and how you will recruit and utilize students in your lab. The key is to demonstrate that your plans for future work are both significant and unique and that the work ties into your previous efforts without simply being an extension of your dissertation or your advisor's research.

7.13.5 The Teaching Statement

The teaching statement serves as a window into the personal philosophy, beliefs, and practice-based experiences of an instructor. It enables prospective employers or colleagues to understand an instructor's goals for student learning, the strategies adopted to teach their subject matter, and the learning environment they create in their classroom. Furthermore, it provides an opportunity for instructors to reflect on past successes and how they have used those experiences to identify challenges beyond the ordinary and build strategies to ensure classes are conducted well.

The teaching statement encapsulates your classroom strategies and is a way to demonstrate to the hiring search committee that you are a good fit for the position. Teaching statements are more important parts of the application process when you are applying for jobs that are more teaching focused. One strategy is to research the institution's teaching mission and encapsulate that into your application.

In my personal experience writing teaching statements for applications, I had more success when tailoring them to the specific position and department. When I received an offer for an on-site interview at St Peter's University in Spring of 2020, I worked on the teaching statement to tailor it specifically to the University's Jesuit mission statement. Of course, I personally knew about the Jesuit's so incorporating this into my teaching statement was not exceptionally difficult, but it was necessary.

In another experience, when I taught entrepreneurship at Stevens Institute of Technology, I tailored my teaching statement to the entrepreneurship department to show my experiences and received an onsite interview and then an offer to teach entrepreneurship for engineers.

7.13.6 Diversity Statement

The diversity statement for professor applications gives you the opportunity to demonstrate how your own beliefs and values align with the mission and vision of the institution. Components include a past experience or an event that helped shape or enhance your understanding of diversity and inclusion, the specific ways in which you have incorporated elements of diversity and equity into your teaching, research, and service, and plans for how you will incorporate them into your future work at that specific university.

7.13.7 Teaching Portfolio

The teaching portfolio includes materials that showcase your expertise as an educator through different documents that help show a comprehensive portfolio of your qualifications, and includes:

- Table of contents,
- Teaching statement,
- Courses taught, workshops, and training,
- Sample syllabi or course plans,
- Sample course materials, and
- Student evaluations (testimonials).

7.14 Interview Process

For those seeking an academic professorship, the job application process is often long and difficult (Office of Career & Professional Development 2012). It is important to apply for many jobs. Many PhDs apply to 50 or more positions during the job application cycle. The first step is to try and obtain as many relevant opportunities as possible. Professor jobs are posted on academic job boards such as:

- Higher education job boards (http://higheredjobs.com),
- Chronicle of higher education (http://chronicle.com),
- Academic keys (http://academickeys.com),
- Inside higher education (www.insidehighered.com),
- Professional organizations,
- Institution-run career pages,
- Twitter, and
- LinkedIn.

Once opportunities have been identified, the next step is to apply with your academic application package detailed above. After successfully applying for a position online, you will likely have to go through early round interviews. These could include one-on-one interviews via Skype with members of the department you are applying to. Finally, if all goes well in the early rounds there may be an offer of an on campus interview. Hiring committees typically invite between three and five interview candidates for on campus interviews, where candidates conduct "chalk talks" to faculty, a presentation to the department, and one-on-one meetings with the hiring department's faculty. The process typically spans approximately two months from the first contact.

7.14.1 Interview Process: Early Round

The early round of an interview process is usually conducted via phone, video conference, or in person at an annual conference. Although it is typically 30 minutes long, the conversation focuses on your teaching, research goals, and interest in the role and institution you are interviewing for. It is important to make sure that you communicate why you are a great fit for the job during this stage of the process so that you can move on to the next step of the interview process. For teaching jobs this might be explaining your relevant teaching experiences and how you fit with the institutional mission. Or for research jobs, this will be talking about your research and the type of work you will conduct.

7.14.2 On-campus Interviews

On campus interviews are a common component of the job search process. For teaching positions, you are required to teach a sample class to a group of students. When I interviewed on campus for a biology professor position at St Peter's University in Jersey City, NJ, I was given the topic well ahead of the interview. I had to present DNA and histone modifications. As a biomedical engineer, I chose to discuss the length of the DNA being approximately 7 ft. in length, and the role of histones in helping condense the DNA in the nucleus. For that one position, they had four on campus visitors and they chose to hire a true biologist by training. Admittedly, being a biomedical engineer, I was not a great fit for the role. During that interview day, my schedule was as follows:

- One-on-one meeting with department chair, faculty #1
- Tour of biology building with faculty #2
- One-on-one meeting with the Dean
- Teaching lesson
- Small campus tour and lunch with department faculty
- One-on-one meeting with faculty #3
- One-on-one meeting with faculty #4

In various campus departments, you might meet with other people whom you might work closely with such as the provost/vice president of academics, human resources representative(s), and president of the university. Each person may have different expectations and aims related to their background. It is important to research the positions of the people you meet to consider some of the questions they might be able to answer for you.

In a research interview, the key component of that interview is the "chalk talk." During the "chalk talk," candidates are asked to present their research using a whiteboard or chalkboard, and then talk about their future plans for grants, collaborations, and hiring PhD students and postdocs (Figure 7.2).

Figure 7.2 The "chalk talk" is a key part of the on-site interview process for a research professor position, where you get to showcase your research interests and plans.

Research faculty positions have similar full day schedules that typically conclude with a dinner with faculty or a tour of the surrounding areas by a faculty member. During this visit, the faculty really wants to give you the best impression of the area so that if they choose you and make an offer, you will be ready to accept. Your main goals during a visit are to stick to your research focus, show your passion for your research and working with others, and maintain your focus on envisioning yourself at that institution.

7.15 Conclusions

Which type of academic institution might interest you? Jobs at the different universities and institutions have different focuses and requirements. Take a look at some of the descriptions in the table below to see what type of institution might interest you (Table 7.2).

In conclusion, applying for and obtaining an academic professor job is a challenging yet rewarding endeavor. It requires a great deal of dedication and perseverance to succeed but it can pay off in many ways that extend beyond financially. Being able to teach or research and engage with students from around the world

Table 7.2 Here are descriptions and some of the common requirements for jobs at each of the three types of institutions.

Doctoral Universities (R1, R2, D/PU)	Master's Colleges and Universities (M1, M2, M3)	Baccalaureate Colleges (Arts & Sciences Focus, diverse fields)
• High research activity. • Requirements include: grant writing, paper writing, publications, mentoring PhDs, designing experiments, and maintaining a research laboratory.	• Teaching and research activities. • Less research focused than doctoral universities with lower publication requirements. • Requirements include: advising, teaching, grading, lesson planning, research, and course designing.	• Teaching focused. • Light research requirements. • Requirements include: advising, teaching, grading, lesson planning, research, committee representations, and course designing.

on stimulating topics can be incredibly fulfilling. Additionally, having the chance to work with other intellectuals in your field and lead research for groundbreaking innovations is highly rewarding. Professor jobs are competitive each year, but as one continues to apply and pursue their passions, success is not too far away – it just takes determination and hard work.

Chapter 7 Key Takeaways

❖ The parts of the application include the cover letter, the CV, the list of references, the research statement, and the teaching statement.

❖ Different job postings and review committees place different weights on different parts of the application.

❖ Tenure review committees evaluate the professor's tenure package at key milestones of the tenure review process.

❖ Although there are outside forces affecting the likelihood of receiving tenure, the likelihood of receiving tenure is almost 100% for professor's who receive grant funding.

❖ Research professor job search committees are more interested in the research statement and plan for receiving grant funding than the teaching statement.

❖ By contrast, the search committees for teaching positions are more interested in the teaching statement and how that individual fits in with their department.

References

August A. (2022). A short guide to the tenure process. Faculty Development – Cornell University. http://www.facultydevelopment.cornell.edu/faculty-resources/ tenure-and-promotion/short-guide-to-the-tenure-process (accessed 11 June 2023).

Bhosale U. (2022). 5 simple tips for writing a good research statement for a faculty position. Enago Academy. http://www.enago.com/academy/5-tips-writing-research-statement-faculty-position (accessed 23 May 2023).

Bloch RJ. (2010). Starting off as a tenure-track assistant professor in a school of medicine. National Institutes of Health (.gov). https://www.training.nih.gov/_assets/slides_3_23_10 (accessed 25 May 2023).

Career Services. (2020). Cover letters for faculty job applications. http://www.careerservices.upenn.edu/application-materials-for-the-faculty-job-search/cover-letters-for-faculty-job-applications (accessed 11 June 2023).

Indeed Editorial Team. (2023). How to write a CV for PhD application (with example). Indeed Career Guide. https://shorturl.at/hoHU1 (accessed 11 June 2023).

Office of Career & Professional Development. (2012). Applying for faculty positions. University of California, San Francisco. https://career.ucsf.edu/gsp/job-search/faculty/applying#Step-1-Know-your-strengths-and-weaknesses-as-a-faculty-candidate (accessed 23 April 2023).

Transition Story: Antonio Marzio, PhD

PhD: Sapienza Università di Roma, Italy
Field of study: Biology
First position out of PhD: NYU Langone Medical Center
Current career position: Tenure track Assistant Professor at Weill Cornell

Motivation for pursuing a PhD: I was born in Italy and transitioned to the United States during my PhD. In Italy, after high school, a Bachelor's degree takes three years and then in order to do a PhD, you must complete a Master's degree. I come from a family of biologists and my family has a biological sciences diagnostics lab that they run. After I finished my undergrad at Sapienza Università di Roma, I specialized in a Master's in biology at the same institution. At this point in my training, I thought that I would become a professor.

After my Master's degree, I decided to remain at Sapienza Università di Roma for a PhD. At this time in my collegial training, I had not practiced much of the biology that I had learned during my Bachelor's and Master's degree. After passing the tests to get into the PhD program, I was finally admitted into a well-funded lab and was finally able to do experiments during my PhD. I was thrilled and loved everything about it. In Italy, you are not expected to publish, and the PhD takes three years with an exam at the end of each year.

Advisor, project, and environment: In Italy, the system is more hierarchical than in the United States. At my university, a full professor would manage a few associate professors and then those associate professors would manage postdocs and PhD students. The full professor could be compared to the head of a business. My principal investigator (PI) was great and well-funded and very supportive of my career. My project was on drosophila and in many ways it was good research but, as stated, without the requirement to publish (like in the United States).

How to Make Your PhD Work: A Guide for Creating a Career in Science and Engineering,
First Edition. Thomas R. Coughlin.
© 2024 John Wiley & Sons, Inc. Published 2024 by John Wiley & Sons, Inc.

Therefore, I did not do the same results-driven research. My fellow PhDs and I were like a big family and we were very united. We had lunch together all the time and we were always helpful.

Moving to the United States: Despite being excited about the research I was doing, I wanted to do part of my PhD in the United States during my second year, I emailed many professors in the United States and told them I would do research for free. Many professors passed on the opportunity as I did not have a publication. After a lot of emailing, I finally received a response from a faculty member at New York University (NYU) at the Langone Medical Center in Manhattan, NY. My Italian university had a residence in Manhattan and I was able to obtain free housing and live on my Italian monthly stipend. At that time in 2012, the exchange rate was in favor of the Euro and my money went a lot farther in US dollars.

At the beginning, my idea was to stay six months at NYU Langone, but I continued for a total of 18 months, finishing my PhD there. Finally, after flying back to Italy for my exams, I was invited back to NYU for a postdoc position.

Start of my postdoctoral fellowship: When I started my postdoc, I maintained my original goal to be a professor. My postdoc went by very fast. It felt like it was just a year. In actuality, my postdoc spanned eight years and six months. That is a long postdoc, and people will say that is an extremely long postdoc, but when I was doing it, I was not concerned about the length of my postdoc. I just loved what I was doing, and trusted myself to continue doing research for as long as I loved it.

Publications and lessons learned: After three years in my postdoc, I was ready to publish my first paper. I aimed for Nature and Cell journals, which have the highest impact factor in science. I went back and forth with these journals going through rejection and rebuttal cycles. Finally, after two years, these journals turned me down. It was a really low point. I thought about leaving academics entirely and going into industry.

After some reflection on my first experience with top-tier journals, I realized it was still a success. Of course, I did not want to have so much time wasted, but as I thought more about it, the experience was not for nothing. In the end, I had learned what "good" looked like to these journals and I knew I could be more successful in the future. Plus, I still had a publication in a considerably strong journal.

After sticking with my postdoc and my research, I worked hard and my next publication was published in Cell. This gave me a huge burst of confidence and affirmation.

Career training: While I was going through my postdoc, I took training courses with NYU's robust postdoctoral office. I took courses on the professor-application package and interview process. In another course, I learned the business side of science and how to manage funding totals. In addition, I received training on the grant writing process. These training really helped prepare me for my future in academics.

Applying for tenure track positions: With my papers and training in my rear-view mirror, I was ready to start applying for tenure track faculty positions. Despite the COVID-19 pandemic in full flux, I was able to apply to professorships during this process. I limited my search for positions to Philadelphia, San Francisco, Boston, New York City, and Houston.

To be honest, while I was looking for a position, I really thought of two questions: where would I want to live for the next ten years and what institution would give me the right resources for good research. I interviewed at a few places including MD Anderson and finally, Weill Cornell Medicine. I was overjoyed to stay in Manhattan and started the position as tenure track faculty in November 2022.

Advice for future PhDs: To someone who wants to be a professor, I would say that the nice answer is to follow your dreams, but more so than that it is equally important to check in with yourself and do what makes you happy and supports your livelihood. For me, I was always excited to do science and check my results.

In addition, I would also tell a PhD students not to compare themselves with others. For example, when my first paper was not published for two years, there was a labmate of mine who had a paper accepted in Cell. I could have been upset about this at that moment, but when he got the notice of acceptance my lab and I went and bought a cake and champagne for him. We all celebrated and the news was fresh air for all of us. The lesson here for me was to take the experience of other people as a demonstration of what was possible and to not get too influenced by other people and to keep moving toward a goal.

Key Insights:

- Theodore Roosevelt once said, "comparison is the thief of joy." During a PhD, it is vital to not compare yourself to others too much and make sure you are personally satisfied with your decisions and career.
- Many times it is important to not look around too much and it is more important to reflect with your own thoughts.

Transition Story: Ada Weinstock, PhD

PhD: Weizmann Institute, Israel
Field of study: Immunology
First position out of PhD: postdoc at NYU Langone Medical Center
Current career position: Assistant Professor, The University of Chicago Pritzker School of Medicine

Motivation for pursuing a PhD: It was my love of biology and immunology that led me to transition into a Master's degree in Immunology and Stem Cell Research at the Weizmann Institute in Jerusalem, Israel.

At the start of my Master's degree at the Weizmann Institute, I rotated through three laboratories. My decision of which principal investigator to work with was easily reached. I focused on working in a laboratory that allowed me to study immunology. Balancing coursework and research, my Master's degree was pretty enjoyable. The people in the program and my exposure to interesting fields made the years go by quickly. With these great experiences behind me, and a desire to delve deeper into immunology, I decided to stay on with my advisor for a PhD.

Advisor: By the time I joined my advisor's lab he was tenured and had made a large contribution to the field. He had an extremely positive attitude and was a very cool person to be around. I was very happy as he supported my research interests and was supportive of my career interests. As I progressed into my second year, my PhD was going well, but there was soon to be a disturbance to the equilibrium.

At the end of my second year, my PhD advisor let me know that he would be retiring at the end of my second year. Soon after his announcement, my advisor committed less and less time to the departmental labs. When he officially retired, I was the only remaining graduate student.

Project: I remember being unsure how the sudden loss of oversight would affect my work. I knew I would need to more closely manage the direction of my

How to Make Your PhD Work: A Guide for Creating a Career in Science and Engineering,
First Edition. Thomas R. Coughlin.
© 2024 John Wiley & Sons, Inc. Published 2024 by John Wiley & Sons, Inc.

day-to-day research activities. My project did not suffer from the loss of my advisor's presence, and I was able to still get research finished in order to wrap up my thesis. Despite the change, I was still able to develop as a scientist from the work.

Environment: The environment was great at the Weizmann Institute. There was a collaborative and dynamic energy among lab groups and the professors. And despite the fact that my advisor left, I enjoyed some of the autonomy and scientific freedom. However, when my third year came around, I was ready to graduate and move on with my career.

Deciding on my next steps: During my PhD I had expressed to my advisor that I wanted to travel outside of the country. After I defended my PhD, my advisor encouraged me to pursue a postdoctoral fellowship abroad. I liked this idea even though I was not set on being a principal investigator as a future career direction.

My postdoc search methodology: I had heard of situations of long postdocs that drag on and I did not want to do a 10 year postdoc to get one publication. So when I applied for postdoctoral positions I decided to create a system for finding a postdoc. I spoke to professors at the Weizmann Institute and asked them how to go about finding a postdoc. My system was as follows: I measured the average time it took to get one publication. I made a list of 80 principal investigator's I could work with in immunology in the United States. Then, I measured how many papers the lab produced over a certain amount of time compared to the number of postdocs in the lab. Sometimes I had seen five Nature, Cell, or Science papers, but there would be 50 trainees. I knew that I would need better odds and a higher frequency of publication per trainee in a PI's lab.

Finally, I dialed back my search for a postdoc advisor and found that I could narrow the advisor list to eight PIs that met my criteria. I interviewed four out of seven labs remotely, and I visited all of these labs on a single visit to the United States funded collectively by the groups. At the end of the interview process, I had four offers, with two of them being in New York City. When choosing between the two universities, the decision factor for me was the amount of funding available. I chose a postdoc at NYU where I would have funding for five years while in the other lab at the other institution I would only have guaranteed for two years.

Start of my postdoc: At NYU, my lab focused on cardiovascular disease. By changing fields from immunology, the skills that I had from my PhD research became very valuable and I was able to bring a fresh new perspective.

Overall, my postdoc was a great experience. My advisor was well-known and had ample funding for our projects. Although he traveled a lot, I was used to being very independent from my PhD experience. At NYU, we were a group of approximately

17 people and had biweekly small group meetings with two or three people. My advisor was involved in the project but did not micromanage anything.

With all my scrutiny and painstaking analysis of postdoc positions, I learned that I did indeed miss a few things. I discovered that the shortcoming of the system I used to find my postdoc was in my assessment of the speed of publications. In my postdoc lab, the average publication speed was every two years. However, my advisor was not very pressed to help us publish quickly. There was a high number of publications, but they could have come out quicker and postdocs might have been able to move through their positions faster. It sometimes took years after leaving the lab to get your work published. In retrospect, I wished I had specifically asked people in the lab about this aspect. From these conversations, I might have learned how often people were publishing and how much help they were getting.

My postdoc spanned seven years and my time during my postdoc included the height of the COVID-19 pandemic. At NYU, they had received funding to support postdocs, and I decided to become the head of the Postdoctoral Council at NYU. In this position, I helped run events for fostering a better environment for postdocs. I also heard many stories from other postdocs who were not in favorable situations. I learned of toxic lab settings, poor leadership from PIs, and a lack of funding. Learning from others and supporting them during their degree gave me a lot of perspective for what I would want to provide my students in an eventual future academic leadership position.

While going through my postdoc, I decided to pursue a career in academics. I began to pursue my career in academics at the end of my third year when I applied for a K99/R00 award. For this award, you can only apply within four years from the date you receive your PhD. I really cut it close! I applied for the grant within two days of my four-year deadline.

I had a borderline score for the K99/R00 grant and I doubted that I would be awarded the grant but then I spoke to the funding officer and they said I will probably be funded. However, while waiting for the grant, the COVID-19 pandemic started and I was unsure if I would still receive it. I thought that most of the funding would transfer to research in COVID-19. After the ups and downs of waiting for the news of receiving this grant, I was finally awarded it! The news changed my whole postdoc.

The K99/R00 gave me a set timeline to begin to set deadlines for my advisor and his reviews of my work. The mentored K99 phase of the award lasts one to two years, and then the grant transitions to the R00 phase where you have to work as a principal investigator at a research institution, and that lasts up to three years.

Applying for jobs: I prepared to apply to jobs in the winter of 2020. However, there were barely any relevant academic job openings because of COVID-19. The job market was abysmal that year. No universities were hiring.

While waiting for the next job market opening period, I spent a lot of time on my application package materials and went to a lot of networking events. I joined the "Future PI Slack Channel," which connected me to an online community of postdocs who wanted to be future PIs. From these resources, I was able to get feedback on my professor's application and fine-tune it to make it stand out.

In 2021, I applied to the professor jobs again. I was not very picky on where I applied because it was not clear which universities would still be hiring during the pandemic despite the posting jobs. It was not out of the question for a department to not hire and instead just test the application pool as they were still adjusting their budgets for the future years from the fall-out from the COVID-19 pandemic.

After the interview process, I received an offer from The University of Chicago Pritzker School of Medicine. I started in June 2022 and since that time have been setting up my lab. My research niche also stood out. I believe that my research is highly relatable and that most people can understand it. In reality, I can collaborate with anyone so that likely made me more easily competitive in the market as well.

Advice for future PhDs aiming to get jobs in academics: For PhDs who are trying to get jobs in academics, I would say that communication skills are very important. In addition to that, my application package was very strong and solid. I had a common thread that tied my research statement, teaching statement, CV, and cover letter together. If a member of the search committee was reading my CV, they would still learn about my research pursuits. Similarly, if they were reading my cover letter, they would still know how many key publications I had and what I wanted to research, and why. I made it very consistent. I believe this helped me stand out. I worked hard on clarifying exactly what I would do in the next steps of my career so the hiring committee could know what they were getting.

I think that above all, choosing a supportive mentor and PI is the most important part of the PhD and postdoc process. If you have a supportive mentor, the science will be more interesting. By contrast, if the lab or group is competitive, then that is on the PI. The PI's tone, and leadership dictates how the lab will be for the PhD students. And no matter what, this is the most important part.

Key Insights:

- When selecting an advisor – if you have a choice – it is important to consider how supportive and present this supervisor will be during your PhD or postdoc.
- While searching for a postdoc position, consider your goals and find out if they align with the lab or group's goals that you are interested in.
- To be competitive for academic jobs, it is important to form or join support groups with other PhDs applying for professor jobs.

Transition Story: John Ruppert, PhD

PhD: Rutgers University
Field of study: Ecology and Evolution
First position out of PhD: Assistant Professor, St Peter's University

Motivation for pursuing a PhD: After completing my undergraduate degree in genetic engineering at Rutgers University, I found myself at a crossroads. I was looking for jobs in industry, but I did not find any jobs that seemed entry-level. Everything required several years of experience, but I was just graduating. I had worked as an undergraduate researcher for three years but had not thought I qualified for more opportunities. I was feeling lost and disappointed. During this time of uncertainty, the program director said I could go for a PhD and the university would fully fund it. I never envisioned myself going to graduate school, but the prospect of having my PhD paid for was too good to pass up, so I eagerly pursued this option. After switching gears to pursuing a PhD, I was able to secure a PhD position with a professor of ecology where I could combine my passion for plants, agriculture, and the environment.

An early lesson: The summer before starting my PhD, I conducted a graduate assistantship in a lab shared by a tenure track professor. One of my first experiences in this lab was witnessing this professor be denied tenure and asked to leave the university. I was shocked. This person had significant contributions with publication in prestigious journals like Nature and Science and yet was denied tenure due to insufficient grant funding. This event made me question the stability of the academic world and contemplate whether pursuing a Master's degree and leaving academia altogether might be a safer path than committing to a PhD. This experience became a pivotal moment in my life, teaching me an important lesson about the unpredictable nature of the academic environment and the need to adapt and persevere.

How to Make Your PhD Work: A Guide for Creating a Career in Science and Engineering,
First Edition. Thomas R. Coughlin.
© 2024 John Wiley & Sons, Inc. Published 2024 by John Wiley & Sons, Inc.

Project: The first two years of my PhD were mostly coursework. In one of the courses, I began to plan my personal research and was reinvigorated about the work. I was working with a private funding source that, in between my second and third year, went bankrupt, discontinuing my project. Simultaneously, I was taking a course about rethinking ecology education. After my undergrad experience, I noticed there was a disconnection between theoretical and applicable coursework. I thought perhaps I could teach and learn how to be a forward-thinking teacher while bringing new concepts into the classroom. The education professor accepted me into a new graduate assistantship, where I worked with a large team, designing modular learning activities for middle school science, and working with and mentoring teachers.

Advisor: My new advisor and I shared a great passion for improving education and I really liked working with her. We really connected on our mission.

Environment: I loved my team. Although my advisor did not have tenure, my advisor's group was a part of a larger group with two other tenured advisors. The combination of professors and other students working in similar fields made for a great environment.

Career Support: Along the entire PhD, I was hesitant to apply to tenure track faculty positions. Instead, I decided I would do a full certification for teaching and get a high school certification. I switched from a research assistantship-supported PhD to having a teaching assistantship and started gaining experience in the classroom.

While teaching, New Jersey's governor pushed for major funding cuts to education, and Rutgers University had to cut teaching assistantships. At this time, I had no funding, had not finished the PhD, and needed a job. I was certified to teach middle school science, and I applied to every open position I could find. Finally, I was offered two positions, a middle school science teacher position and a lab manager and instructor position at St Peter's University. Choosing between these two roles was a very hard decision. Teaching middle school paid better, but after some discernment, I chose to pursue the position with St. Peter's University. I was excited to have some undergraduate researchers to mentor and pursue research with.

I completed my PhD in 2015 while keeping my appointment with St. Peter's University. Then, a year or two afterward, one of the faculty members at St. Peter's University retired, and I applied for the open position. To my delight, I was offered the tenure track assistant professor position.

Looking back: Having a tenure track job is very secure and offers me a lot of flexibility. I do not report to a manager. Instead, I have a tenure review committee that reviews my progress, but no one is telling me exactly what to research or how to design my courses. The research is my direction, and I absolutely love the independence and ability to pursue my interests.

Advice for future PhDs: I think it is natural to want to go with the flow, but that is what everyone does. For me, that experience before my PhD where I saw a professor, who I admired, be denied tenure stayed with me for the duration of my career. That moment made me anxious and that insecurity, I believe, helped me truly understand the realistic view of academics. Academics are not always secure. You can lose funding, universities can lose grants, and government politics can affect budgets. During my time as an academic, I learned that if I focus on my career and my work, and do not get too involved in some of the politics that can be present at some institutions, I could be successful. Being confident in your abilities and sticking to what you know is really important. Academics can feel like an insecure place, but that does not have to be your demise. Instead, I would advise you to be proactive and let that insecurity fuel you to stay ahead of job security and prospects by learning how to navigate it.

Academic job applications: There is a business to academics, and institutions do not create jobs they do not think they will be able to support. And right now, with Generation Z being smaller in size than the millennial generation, universities want to make sure they are not oversized for the population. Right now, it is hard to see academics as a secure job market because there is a lot of focus on adjuncts. But when applying for academic jobs, make sure you read the mission statement of the university, write a good cover letter, and make sure to make each application personal to the department you are applying to.

Key Insights:

- It is easy to simply go with the flow, but Dr Ruppert learned early on that it is important to carve your own path and be your own guide in life, and especially in the PhD job market.
- With decreasing rates of enrollment with Generation Z and more universities opting toward using adjuncts for teaching positions, there are likely to be fewer available tenure track positions in the decade of the 2020s.
- In order to navigate this challenging landscape, it is important to focus on what you can control by giving yourself the best opportunities for success (e.g. writing grants, honing skills, creating great job applications, etc.).

Part III

Your Nonacademic Path

8

Nonacademic Careers

8.1 What Careers Are Available for PhDs?

Until you truly start to look for PhD careers outside of academics, they will be a vague, misunderstood concept. But once you begin to look for them and find them you will discover that there are many fulfilling careers for science and engineering PhDs.

In this chapter, you will see how academics and industry are intertwined on the path of discovery to development. Inside each step along the path from discovery to development, there are numerous jobs for PhDs that are described in this chapter (Figure 8.1).

Some of you may have heard the phrase, "What can I be with a PhD?" That is because it is the name of the weekend course hosted once each year at New York University (What Can You Be With a PhD? 2023). This event aims to expose PhDs to a number of job opportunities outside of academics. It is an amazing two-day event where PhDs in all sorts of careers talk about their transition and job role. Unfortunately, there are not many events like this one, but fortunately, I went to this event and interviewed numerous PhDs about their job prospects and career choices (Coughlin 2021).

Utilizing insights gained from the What Can You Be With A PhD (WCUB), interviews with PhDs, strategies gained from alternate sources (Cheeky Scientist 2023), research with National Institutes of Health (NIH) reporter, and PhD source surveys, this chapter gives a detailed overview of nonacademic career paths available to PhDs and how they contribute to the innovation landscape.

How to Make Your PhD Work: A Guide for Creating a Career in Science and Engineering, First Edition. Thomas R. Coughlin.

Figure 8.1 Congratulations on opening the door to your post PhD career plans. Every step is one step closer to finding what you are looking for.

8.2 Visualizing Jobs on the Path of Discovery to Implementation

In science, if you look at the US science and engineering economy, it can be viewed as a tree. The seed of the tree is discoveries, which normally occur in academics, and the tree that grows from it is the life cycle of that discovery. There are countless jobs to support the life cycle of a discovery and countless companies that make their living off of moving discoveries through our government's and country's processes of approval before the discovery and product are able to be distributed, marketed, and sold to a consumer.

At a glance, there are jobs in government, nonprofits, academia, startups and venture capital, finance, and industry (Figure 8.2). All of these different sectors help bring innovations to the American consumer. And many of these innovations start their path in academia, where they are funded, discovered by principal investigators, patented with the help of technology transfer and commercialization departments in academics, licensed to industry or converted to a startup, and then created into a product. These smaller startup companies eventually get acquired by bigger companies or become bigger by creating more innovations.

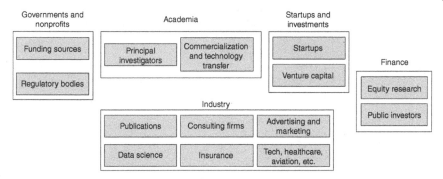

Figure 8.2 This diagram shows the different sectors of the economy that a PhD can work. The areas are broken up further into specific groups, careers, agencies, or companies.

Industry is built to support itself through unique synergies between specialty industries while being invested into by public investors, and venture funds. I will explore each of these areas of the economy with the PhD jobs in this chapter.

8.3 Publications

With a specific idea, a researcher is able to write a grant proposal or application to fund that project. Then, the research process begins, taking ideas from identification of a problem or potential mechanism to reviewing the literature, setting objectives or hypotheses, study design, sample design, data collection, processing and analyzing data, and reporting that data through publishing. Academics publish their findings with the help of publishing companies and journals, enabling the dissemination of their findings.

8.3.1 Journal Editor or Senior Editor

Description: Journal Editors or Senior Editors oversee the peer-review and publication process of original content. They take ownership of overseeing submissions and a rigorous review process to ensure publication of high-quality content. They build global relationships with people and institutes in the translational research community to identify new material, authors, subject experts, and content opportunities.

8.4 Commercialization and Technology Transfer

The academic advisors and their groups also work with the commercialization and technology transfer departments within their institutions to protect their research and create intellectual property (IP) (Figure 8.3). Through this arm of

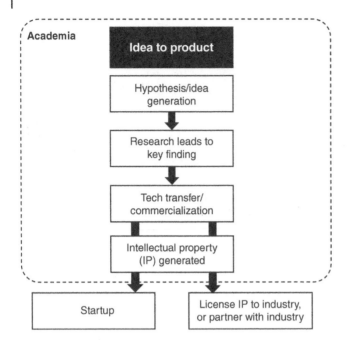

Figure 8.3 Academic institutions hire PhDs in their commercialization and technology transfer offices to help create intellectual property with their faculty. The intellectual property can then be spun out into a startup or licensed to industry.

the institutions, the academic researchers can protect their work and also license their work to startups or industry through working with business development offices in startups or industrial companies. By having commercialization and technology transfer offices, institutions are able to create IP and enable monetary gains from these innovations or inventions. Institutions will either license their IP to industry companies or incubate their IP in the form of startups that are formed and housed in their respective institutions. In most cases, the institutions retain a percentage ownership of the IP to earn royalties from successes of their academic researchers. This way the institutions foster the transition from idea to research to company creation, fostering a burgeoning intellectual environment.

8.4.1 Intellectual Property Liaison (Also Called: Licensing Manager or Technology Transfer Officer)

Description: Intellectual property liaisons partner with researchers and legal professionals in the creation, management, and protection of IP assets.

8.4.2 Innovation and Commercialization Manager

Description: On the business side, the innovation and commercialization manager plays a critical role in the success of new technologies by supporting collaboration and licensing activities aligned with the research and development strategy.

8.5 Startup Scaling

Many academic institutions have entrepreneurial labs, also called startup incubators. It is advantageous for academic institutions to have startup incubators for students to create companies for several reasons, including facilitating new experiences for undergraduate students and fostering commercialization of their university technologies. Furthermore, as entrepreneurship is a common term because of television shows like "Shark Tank," academic institutions have these startup incubators to attract students and foster innovation. Directly, startups in academics help them gain prestige.

A startup is an exciting and challenging endeavor, with no guarantee of success. The first step is typically to put together a comprehensive business plan that outlines goals, strategy, marketing tactics, and how financials will be handled. Once the initial plan is set in place, it is important to start cultivating a network of advisors and venture capitalists who will provide guidance and resources as the company progresses. Following that, entrepreneurs must look for possible investors or customers through targeted campaigns, emerging markets, or strategically positioning their product in competitive marketplaces. With hard work and perseverance, a young startup can slowly but surely cultivate a customer base and eventually become a profitable venture.

There are many funding sources available to startups including institution awards, grant funding, and competitions. After the initial seed stages of a company, venture capital (VC) investors may step in and provide series A or series B funding to the startups. Startups have a high probability of failure yet are an essential part of the economy and innovation landscape.

8.5.1 Pre-startup Stage

Before the startup is truly a company, the startup is an idea. While a startup is an idea the founding team members conduct market research, gain customer insights, and hone their product design to fit their defined customers' needs. From there, the team moves into concepting and prototyping. During these early pre-startup stages of the company, they are too early for seed level or Series A or B investing. Instead, these early stages are typically supported through crowdfunding, family, friends, or the founding members.

8.5.2 Startup Stage

The startup stage of the startup is when the founding team transitions from concepting and prototyping into creating a minimally viable product (MVP). The MVP is a version of the device or product with just enough features to be usable by early customers who can then provide insights and feedback for the future product development. From the early MVPs, the product is refined from these customer insights until the product is no longer in customer testing. From here, the product moves from being a startup into a place where it is ready to scale. In some companies, the graduation from an MVP into a finalized product will mean that they are no longer only ready for seed-level investors like angel investors or family and friends, and are instead ready for venture-level funding (Figure 8.4).

8.5.3 Growth Stage

During the scaling phases of a startup, the company is growing its customer base, increasing production while raising marketing efforts. Growing too quickly can leave startups without enough products or staff to sustain their growth. By contrast, growing too slowly can leave a startup without income to please investors. A lack of scale can leave investors displeased and lead them to step in to change things up in the company executive team.

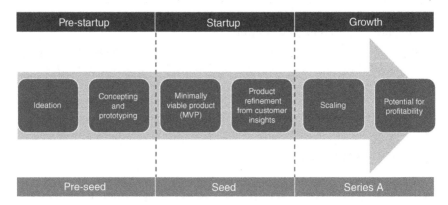

Figure 8.4 There are three stages of a startup where the company moves from pre-startup, into startup, and then into growth. These stages of the company involve specific steps as the founding team moves from ideation to MVP to a final product and into scaling and eventual profitability. During each of these stages, the companies unlock access to different levels of investing that are respective to the stage that the company is in.

8.5.4 Startup Companies Offer Potential High Risk and High Reward

On the receiving end of this innovation paradigm shift are the startup companies. By shouldering more risk, startup companies either fail or succeed. Working at a startup company can be very time-intensive, but the reward can be very high. In addition, working at a successful or eventually failed startup company can sometimes allow you to stretch across roles and gain a wide array of experience that can be very helpful as you grow in your career. Newly minted startups are typically operating on a tighter budget, which provides a lot more diversity in the work that you can perform at them. In fact, it is encouraged to learn how fundraising, grant writing, pitching, and hiring occur in order to be able to speak to these skills in interviews later on in your career.

8.5.5 PhD Level Jobs in Startups

Description: A career at a startup may be in a range of topics. The types of job roles needed at a startup depend on where the company is in its progress. Therefore, there is not one type of job in a startup in which PhDs slot into. Instead, there are multiple jobs that PhDs can take. Many times a PhD might work on the science or engineering of the company's product or service in order to progress the company forward.

8.6 Venture Capital and Startup Growth

There are a few different types of funding available to startups. The fundraising stages of a company are:

1) Seed funding (angel investors)
2) Series A (venture capital investors)
3) Series B (venture capital investors)

At these different stages of funding, the company is invested into by angel or venture capital investors. Before the seed stage of funding the company is considered to be "pre-seed," meaning they have not received investor funding and are operating on their own funds or grants.

The grants that businesses typically apply to are government or incubator competition awards that give businesses early funding and the business does not have to give up any equity. In other words, the company remains to be 100% self-owned.

Depending on the type of company, the pre-seed stage of the company could give them enough capital to be able to create a profitable product. Alternatively, the company may seek additional funds from an angel investor or venture capital group.

At the seed stage, the startup company aims to give up part of the ownership of the company at some cost to an angel investor. There are many ways to structure these deals, but overall the investor will get some percentage of the company and the company will get a lump sum to be used to achieve milestones that are agreed upon by the company and the investor.

Again, there are companies that can utilize their pre-seed and seed funding to create a profitable business. And in these cases, the company could become self-sufficient where they have products that they produce and sell to businesses or consumers, thereby creating some revenue which they can reinvest into future products.

In the case of pharmaceutical drug development, some tech companies, and other companies might need additional funds to carry out their product creation and scaling of the startup to support this drug, device, or app. In these scenarios, the company would seek our venture capital funding during a Series A raise. During this stage, and similar to the angel seed investment stage, the startup would have to sell some of its ownership in order to obtain the capital necessary to carry out the funding steps. The company can also elect to fundraise at a Series B raise, where the company has demonstrated a higher worth and the cost to a venture capital group to buy part of the company is higher.

Overall, as startups demonstrate more and more success by achieving key milestones in their product approval process, the value of the company increases. As such, the investors that invest later have to pay a higher price to obtain part of the company. With that, an investor who invests earlier in the seed stages will own more of the startup company for the same price at later stages of the fundraising stages, but the investor will also take on more risk because the likelihood of failure is higher. And therefore, in the negotiation with the company, the company gives up more ownership to the angel investors to counteract the risk and offer a higher reward for a potential success.

When a startup company has de-risked itself by achieving certain milestones, the company may become attractive to a larger industry to buy the startup. In these scenarios, the larger industry company may elect to acquire the startup or merge with it to form one company.

8.6.1 PhD Level Jobs in Venture Capital

Description: A venture capital consultant evaluates emerging technologies and companies for a venture capital company or investing group. They also conduct technical and market diligence and analyses of the company in order to determine the value the company brings and whether the company's valuation is correct in order to provide investing feedback to the group.

8.7 Mergers and Acquisitions Are a Main Form of Acquiring New Innovations

This might come as a surprise to you, but many industrial companies have moved away from conducting their own research in-house. Instead, large-scale corporations have realized that it is more advantageous to buy companies and their associated IP through acquisitions.

These larger corporations have realized that by acquiring assets from companies they are removing the associated risks associated with the earlier stages of research. Research takes a lot of time, money, and resources, and on top of that, many ideas never making it to market. In the end, if the venture is not successful, it costs the company lots of money without any return on investment (ROI). The leadership at multiple companies have collectively realized that since the likelihood of failure of new startup or new ventures is roughly 90%, it is not worth the cost to have their business take on so much risk. Therefore, large companies are able to mitigate the cost of finding new solutions (which are only successful about 10% of the time) by simply buying the assets of successful startup companies.

One example of a successful acquisition was when Amazon agreed to acquire Whole Foods on June 1, 2017. Amazon executives made an offer of purchasing all of the shares of Whole Foods at a price of $41 per share. Whole Foods countered with $45 per share, and then they accepted an offer from Amazon of $42 per share. By buying all the shares of the company, Amazon owned all of the parts (shares) of Whole Foods. The acquisition gave Amazon the well-known brand of Whole Foods, distribution centers, locations, and the ability to integrate Whole Foods with its online Amazon store. As such, Amazon benefited from having all of the assets of Whole Foods to build into its already well-established company.

In the pharmaceutical industry, large companies like Bristol Myers Squibb, Merck, Pfizer, and Johnson and Johnson have become more accustomed to buying assets from startups. Merck and other companies still keep innovating on products and drugs, but much of this innovation is in the way the drug is marketed rather than in the development of new drugs. As such, the large pharma companies can specialize in running large-scale clinical programs and compete for market space by achieving new indications for their existing drugs that have patents. With patents lasting 16–20 years, the pharma companies have plenty of time to capture market share and achieve global market adoption of their key drugs to achieve full capitalization of their IP.

8.7.1 Acquiring a Successful Startup

As mentioned, in the pharmaceutical industry, acquisitions are very common. Because of the potential losses that can be incurred from failed projects, larger pharmaceutical companies do not do as much in-house research of new innovative technologies as they used to do. Instead of taking these projects on in-house, the company will buy the assets of a startup or company that have already proven to have a successful product. In this way, the larger pharmaceutical companies can combine new drugs or "assets" with their existing portfolio of assets. In doing so the company focuses their resources on increasing market adoption and sales.

8.7.2 PhD Level Jobs in Mergers and Acquisitions

8.7.2.1 Business Development Manager

Description: A business development manager works closely with commercial, marketing, research and development, and operation teams to lead business development efforts of the company. Execute business development initiatives and achieve annual sales targets for territory, while also identifying key prospective clients and perform strategic, scientific, and tailored outreach to increase new business and expand existing accounts within designated territory.

8.7.2.2 Management Consulting

Description: Management consultants help make intelligent, informed decisions for business happen. They work on topics ranging from diligence through strategy and execution, focused on evaluating technology-driven markets, assessing technology companies, and helping technology companies solve their toughest challenges. They also conduct both marketing and technology diligence in order to make informed business decisions.

8.8 Industry Companies

Given that many large companies de-risk by acquiring startups, you might wonder what the work at a large company actually entails. A large company has many roles that center around protecting and capitalizing on its main assets. The main assets of a company will be its products and services. The company will provide either a service or a product and the company staff will work on creating a community that works to sell and collect revenue for that good or service. Simply put a company is like a small operating charter or town that needs to operate smoothly in order to create success and continue to run smoothly.

Every year, Forbes Magazine lists the top 100 companies to work at. Year in and year out, International Business Machines (IBM) makes the top 10 companies to work for in Forbes. The employees at IBM cite good benefits, healthy work-life balance, and a well-organized company structure as the hallmark characteristics that make IBM great to work at.

Figure 8.5 This diagram shows the different departments of a company. At multiple stages of growth, these departments are necessary to operate a company. Larger markets and more complex products require more marketing teams, internal regulatory teams, as well as larger research and development teams and technical support.

Many companies have realized that creating a healthy work culture helps to foster good products and services for the customer. Many CEOs cite making their employees a priority over their customers because happy employees will lead to healthy employee–customer interactions.

This basic diagram shows the different departments of a company (Figure 8.5).

8.8.1 PhD Level Jobs in Large Companies

8.8.1.1 Medical Science Liaison

Description: Medical science liaisons serve as the scientific interface between the company and the healthcare professional communities with respect to communicating scientific information about the company's products, scientific data, and clinical development plans. They are responsible for developing and maintaining relationships within the medical/scientific community and providing medical and/or scientific data about company products and research to healthcare professionals. Their clinical, scientific, and technical expertise will be maintained through review of scientific literature, attendance at assigned medical meetings, and self-learning. The job includes lots of travel and schedule flexibility and autonomy in making your own appointments.

8.8.1.2 Product Sales Specialist

Description: The product sales specialist accelerates revenue and market penetration for all products in the territory while driving the development of relationships with key customers. They conduct key marketing activities such as peer-to-peer events. They execute the sales plan and produce monthly and quarterly updates to the plan and review process.

8.8.1.3 Principal Engineer

Description: The principal engineer utilizes their engineering skill set in a specific and defined position. As a PhD engineer, this position is more defined and specialized and may offer a faster transition into project management and personnel management.

8.8.1.4 Scientist
Description: A scientist utilizes their scientist skill set and critical thinking, experiment planning, data interpretation, and study design. As a PhD scientist, this position also may offer a faster transition into project management and personnel management in a given organization.

8.8.1.5 Marketing Research Analyst
Description: A marketing research analyst may work on a variety of projects, which depends on the level of marketing needed, and the customer size. This position focuses on devising commercial, branded, and non-branded content to educate and promote adoption of the company's key and upcoming products. The key role of the marketing research analyst is to determine the key pain points of the target customers to explain how the existing and upcoming products meet those needs. In addition, the marketing research analyst conducts advisory boards with key leading customers, and communicates across teams to carry out actionable steps that create materials that foster adoption.

8.8.1.6 Business Development Manager
Description: The business development manager utilizes networking, communication, and strong technical skills to foster meaningful partnerships and IP licensing opportunities to create a continued pipeline of products. Key actions in this role include developing a market strategy and building new partnerships.

8.8.1.7 Regulatory Affairs Specialist
Description: The regulatory affairs specialist's roles include drafting and reviewing regulatory submissions (IDE, Q-sub, 510[k], De Novo Requests and PMA submissions) as applicable. Further, this role includes preparation of regulatory submissions in specific geographic regions (many countries have their own regulatory processes), and assisting post-market regulatory activities, domestic and/or overseas. This role can also manage interactions with regulatory authorities, notified bodies, and certification bodies.

8.8.1.8 Technical, Scientific, or Medical Writer
Description: Some companies outsource technical and medical writing to Technical/Promotional or Medical Communications agencies, respectively. However, some companies have an in-house writer. Responsibilities of this role include utilizing expertise and specialty knowledge to conduct technical writing that adheres to regulatory guidelines of the different types of pieces. Work includes promotional, commercial, and educational pieces that require different types of writing and copy to communicate across different audiences.

8.9   Regulatory Agencies and Legal Services

The industry companies' products must obtain approval for their products. Approval for products in different industries means different things. Approval in the pharmaceutical industry means having a drug or therapy approved by the Federal Drug Administration (FDA) or the EORC. By contrast, approval in the medical product industry also needs approval through the FDA but goes through a different track or pipeline. In addition, in the battery industry, auto industry, and other industries there are alternate agencies and groups that require specific regulations, criteria, and/or standards to be met. For example, in the automotive industry, cars need to meet certain emissions standards.

Some companies have their own legal departments, while others hire outside consulting services. Legal services can help with the copyright, patents, and trademarks for the industry product and company. In addition, legal departments will ensure that the IP of the company is protected from being stolen by outside competitors.

8.9.1   PhD Level Jobs in Regulatory

8.9.1.1   Regulatory Affairs Associate/Manager

Description: Regulatory affairs departments draft documents to support interactions with the guideline agencies' requests, regulatory plans, and other regulatory documents as required. They develop and maintain current knowledge of regulations, guidance, and standards applicable to assigned projects. They analyze relevant regulatory information and provide updates to the project team. They assist in developing and maintaining regulatory affairs department procedures and process improvements; and support the preparation of regulatory documentation and submission activities to meet business and agency milestones.

8.10   Sales, Marketing, and Communications

Many companies use outside marketing and communications services to help promote adoption of their goods or services. Their internal team at a company will work with boutique agencies that specialize in one aspect of the sales and marketing process. In order to have a product or service adopted by a business or consumer, that entity or person must see the benefit of that item. Therefore, convincing them of the benefits of the product or service is the job of the sales and marketing teams.

The sales team are experts on the product who also know sales techniques and strategies. The sales team also includes technical support services that can involve PhDs during the education and installation process.

The marketing team conducts market research to determine ways to promote the product. The four P's of marketing include product, price, place, and promotion. Depending on the target market of the product, the marketing team adjusts its budget to focus on reaching their customer. Overall, the team creates awareness of the new product by establishing the customer's trust and attention.

The communications team works hand-in-hand with the marketing team to create the marketing media that is disseminated to the customer. In the medical industry, communications is outsourced, and medical communications agencies work with pharmaceutical companies to conduct market research and develop the educational, publications, and marketing materials to reach the target customer.

8.10.1   PhD Level Jobs in Sales, Marketing, and Communications

8.10.1.1   Product Sales Specialist

Description: The product sales specialist accelerates revenue and market penetration for all products in the territory while driving the development of relationships with key spine surgeon customers. They conduct key marketing activities such as peer-to-peer spine surgeon events and visiting surgeon programs that espouse the values of our foundation technologies. They also, execute the sales plan and produces monthly and quarterly updates to the plan and review process.

8.10.1.2   Agency Technical/Medical Writer

Description: The difference between this role and an in-house technical/medical writer is that an agency receives contracts from companies to work on their products. The array of projects can differ and these roles often require flexing across accounts and learning new areas of work to be able to write on multiple technical and/or therapeutic areas. This job is never lacking in pace or excitement. Similar to an in-house role, the responsibilities of this role include utilizing expertise and specialty knowledge to conduct technical or medical writing that adheres to regulatory guidelines for different types of pieces, including promotional, commercial, and educational pieces, which require different types of writing and copy to communicate across different audiences.

8.11   Investment Banking and Equity Research

Companies traded to the public with stock offerings and investment opportunities are evaluated by banks, hedge funds, and venture fund analysts. At banks and hedge funds, equity research analysts determine the value of a company by analyzing that companies' assets and their market share. Ultimately, investment teams aim to determine if a public company is valued appropriately on the stock exchange.

Equity research analysts work for both buy-side and sell-side firms in the securities industry. They produce research reports, projections, and recommendations concerning companies and stocks. Typically, an equity analyst specializes in a small group of companies in a particular industry or country to develop the high-level expertise necessary to produce accurate projections and recommendations.

These analysts monitor company news, data releases, and marketing updates, while also conducting interviews with corporate executives from the companies they monitor. Using these monitoring techniques, the equity research team updates valuations to determine the impact that these activities have on the price of the stock. The job of equity research can be conducted on either the buy or the sell side of the industry, where the investment company determines whether they should buy a stock in a publicly traded company or sell a stock. A single equity research team evaluates a portfolio of companies.

8.11.1 PhD Level Jobs in Equity Research

8.11.1.1 Equity Research Analyst
Description: An analyst position is an entry-level role into equity research. They analyze industry macro trends and company financials under the direction of the senior analyst. Prepare research reports varying in size and scope from quick briefs on earnings results or strategic events to more in-depth, original thought pieces. They build and maintain financial models and industry databases for evaluating companies and market segments. They review business and trade publications, annual reports, financial filings, and other sources in order to gather, synthesize, and interpret data on companies followed by a team. They research, analyze, and forecast the impact of technology as well as varying corporate strategies. They are responsible for writing research reports. They also communicate recommendations to sales, trading, management teams, institutional investor clients, and other important stakeholders. Finally, they also develop marketing materials for clients and attend industry conferences where appropriate.

8.12 Conclusion

The flow of discovery relies on many jobs, including researchers, technology transfer, publications, startups, venture capital agencies, pharmaceutical or large tech companies, marketing and communications companies, equity research groups, sales, regulatory agencies, funding agencies, and more.

Chapter 8 Key Takeaways

❖ On the path of innovation, from initial discovery all the way to delivering a product to a customer, there are many jobs for PhDs.

❖ Each of these jobs requires different types of skills and might fit different types of people. In addition, these jobs might offer more monetary gain with more potential for advancement while others might offer more security.

❖ Startups offer the ability to learn quickly in a fast-paced environment with more risk but higher rewards. By contrast, industry jobs in more established companies may provide more security and immediate financial reward without a potential for payoff.

References

Cheeky Scientist. (2023). Cheeky scientist gets PhDs hired into industry careers. Cheeky Scientist. www.cheekyscientist.com (accessed 15 May 2023).

Coughlin, Thomas. (2021). My take home messages from the WCUB – shrinking academic job market creates new avenues for PhDs in America – PhD source. PhD Source. https://www.phdsource.com/blog/my-take-home-messages-from-the-wcub-shrinking-academic-job-market-creates-new-avenues-for-phds-in-america (accessed 15 May 2023).

What Can You Be With a PhD? (2023). WCUB2023. www.whatcanyoubewithaphd.com (accessed 15 May 2023).

9

The Industry Mindset

9.1 Industry is Not Academics

Industry is different from academics. You have likely heard this before. But why? In this chapter, I will discuss the hard lessons I learned from industry, and how industry is different from academics.

9.2 Industry Lesson #1: Expendable

When I first transitioned from my postdoc to industry, I was 29 years old and starting a job that I never thought I would do. I could not imagine having a traditional office job, sitting at a desk. The office with the other workers – now my colleagues – was an alien environment and concept to me. I wondered, how could all of these people possibly want to do this? And how could they find any fulfillment from carrying out someone else's vision, every single day? My mindset was strictly academic. Despite my education and coursework at New York University (NYU) during my postdoc about working in industry, I was still looking at things from an academic perspective.

Despite these lingering thoughts, I did my best to fit in as a medical writer at the busiest pharmaceutical medical communications company in New York City, if not the entire country. I was eager to understand the culture before determining if I wanted to fit in with it. I decided I was going to figure out how the industry worked. I wanted to understand the different departments, how people led, and how the work got done.

I made amends with my job by focusing on what I enjoyed. I liked being paid to stay on top of a field that I had been strictly focused on the earliest mechanisms of action and research. It was exciting. I saw my career as spanning from

How to Make Your PhD Work: A Guide for Creating a Career in Science and Engineering, First Edition. Thomas R. Coughlin.

understanding discovery in academics at the bench to reaching the customer in pharmaceutical drug development. I felt this was truly the real perspective of how it all worked that I described in Chapter 8.

After a positive three-month review from my manager and then moving to another group within the company, my stint at my first position in industry came to an abrupt end. In one day, the company lost 20% of its revenue from its main client, Bristol Myers Squibb. To make up for the losses on its balance sheet, the company laid off 25% of its workforce. I remember leaving that day with my other colleagues who had also been fairly new at the company, and I was not the only PhD.

After the layoff, I second-guessed my career decision to go into industry. "How could I go through this later in my career?" I asked myself. "What if I had a family and got laid off?" These questions circulated through my head. This was the first time I really had the feeling that despite all of my effort getting to know my colleagues, making relationships with them, and even coming up with some ideas for writing some code for the company, I was, in fact, expendable. I did not bring in money. My entry-level position was just that, an entry-level position without much leverage. I did not have any established relationships with clients, and I was not going to be noticed when I was fired. I could be replaced with another PhD. It was the singular most important lesson I received in industry. And one I will not forget.

9.3 The Role of a PhD in Industry

Your job in industry is to help the company make money. Initially, your entry-level role is to simply do your job (Figure 9.1). The hiring team brought you on to do your work as reliably as you demonstrated you could during your interview and on your resume. Anything in your job description is your responsibility (Cannon 2016). Sometimes even tasks that are not in your job description will become your responsibility, too. Promotions will come, but only after you truly understand your role in maintaining the core elements of your organization running effectively and efficiently.

To move up in a company, you must understand, what makes the company make money? And then, you must figure out how to lead the people in the company with you to achieve those goals. How can you carry out those actions and tasks to motivate, inspire, and lead those around you?

9.3.1 Trust is Key in Industry

In all companies trust is the number one reason a company makes money. Whether it is trust in a product, a relationship, or an expertise that you are paying for, a customer's trust in a product or group of people is the reason a company makes money (Wong 2020).

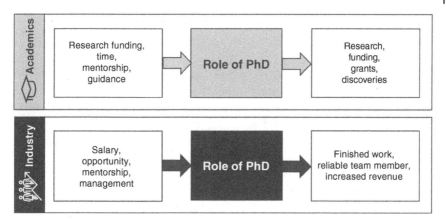

Figure 9.1 PhDs do not have the same motivations or performance goals in academics and industry. They are different. PhDs in academics spend research funding, time, receiving guidance and mentorship in order to contribute to research, obtain funding, get grants, and make discoveries. By contrast, a PhD in industry receives salary, a job opportunity, some mentorship, and management to be able to complete assignments, be reliable, and increase revenue of the company.

In medical communications, for example, trust in relationships between an agency and an industry collaborator is the primary reason that a company entrusts an agency with its work. Cultivating reliability and accountability, with strong products helps the client build trust and secures the relationship. Finding how to build trust in your role by contributing to the company's value is a good way to start being a trusted member of your team.

9.4 Academics is Like a Business

As a PhD in academics, we are not trained to understand academics from the viewpoint of a business. But a research group or lab is, in fact, a business. The grant money is only given to the supervisor if they can demonstrate the right ability, prestige, and accountability to get the work they outlined in their grant completed. The grant officials must be able to trust that the supervisor and their lab or group will get the work done.

A research supervisor builds prestige in a given field with publications and conference presentations. Then, using this prestige, they secure funding by writing award-winning grants. This cycle in academics relies on money but is more driven by prestige, ideas, and researching grant topics that are interesting to funding agencies.

9.5 The Main Difference Between Academics and Industry

The main difference between academics and industry is that the research supervisor and their group do not generate income for their research unless they receive more grants (Figure 9.2). Businesses must have income above their operating costs to survive. Of course, a business can operate in a state of debt for some time, but eventually, it must have some profitability. By contrast, an academic group or lab must have some grant funding to stay afloat.

In academics, publications are the main product of the research group, which appeals to grant review boards to obtain more money. In this way, the academic supervisor's group becomes reliant on funding sources to be awarded money to conduct their work.

The main alternative way for a supervisor to receive money is to have a contract or licensing agreement with an industry company or government contract. Business will pay for academic institutions to consult for them on projects, conduct

Success metrics of a research group or company

Figure 9.2 Companies first start with capital from sources like individuals, private equity, and then venture capital. But after the early stages of a startup and the company starts selling products, the company must remain profitable to stay in business. In order to do this, they must sell products, have market capitalization, and grow in customer loyalty and trust to stay in business. By contrast, the success metrics for academics are obtaining grant funding, having more publications, publishing in journals with high-impact factors, increasing citations, and presenting at conferences. Most of all the success metrics of academics and industry are motivated by slightly different metrics. And the trust needed to succeed in academics comes from being able to get more grant funding, while in industry the trust comes from customers and being able to maintain sales and increase them.

innovative work, or answer challenging questions. In addition, a research group might also create intellectual property as you learned in Chapter 8, where the intellectual property is sold to industry through licensing agreements or purchases.

9.6   Industry Lesson #2: Customer Relationships and Risk

After being a part of two startups in financial technology and healthcare, I decided to start my own company. In June 2020, right at the start of the COVID-19 pandemic, I started my own business helping small businesses adopt new technology for their business and consulting for startups. The timing was challenging to convince small business owners that they could increase revenue by spending a little more on the right technology to increase sales, increase margins, and drive more customers to their websites. After about a year, I had grown my monthly recurring clients from six to twelve and was making enough to have a self-sustaining business. But then I learned a valuable lesson. I lost a key client. I had not called that client in two months and was relying on them to not seek another service provider. I was wrong. When I finally did call the customer, they had already switched to someone else. I learned very quickly that keeping in touch with client's and asking about their satisfaction is of utmost importance. And that truly my livelihood depended on these relationships. Knowing those were stakes, I would never forget that those relationships mean life in business. Furthermore, I also learned that if my business was so affected by losing one client, that I needed to de-risk and distribute that risk around more clients and more areas.

These two lessons on the importance of relationships and the importance of spreading around risk, helped me understand the vital role that customers play in a business' life. Without them, you fail. And if you neglect clients and do not have the right stop gates in place, you lose revenue and your entire business can sink. I was finally on the other side of the layoff I had experienced in my first job. I realized that employees are important, but the bottom line is ultimately always more important. And anything that jeopardizes a relationship with a customer base is the most critical of all. Damage the trust of a customer or lose their interest, and you will not be able to stay afloat. Customers and trust are life in business.

9.7   Industry Lesson #3: Align Yourself With the Company

The third most important lesson in industry came when I transitioned back from startups into industry. At this point in my career, I had taught entrepreneurship in startups and been a founding member of two startup companies. When at those startups, every single person on the founding team is aware of the bottom line of the company, and when we hired more people, we had to hire people who were on board with doing their job well. It is important to carry out the values of the company and understand the work you do. Of course, fulfillment comes from the overarching mission, but it is important to understand the main purpose and money-making part of the business. The lesson of aligning yourself with the goals of the company allows you to be an accountable person in a company for your colleagues and management to rely on.

9.8 Conclusion

Transitioning from the world of academia to industry can be a challenging experience. The learnings in this chapter and the mindset it takes to succeed in industry are focused on the bottom line. Knowing why your job is critical to a company is essential for your own personal growth and development. Beyond just mastering new skills, it is important to understand how a company runs and makes money. The main takeaways are:

1) In industry jobs are expendable and the bottom line is the focus,
2) Trust and customers are life in business,
3) If you align yourself with the company's goals, you will achieve success in industry.

To put it simply, once all these lessons have been learned, a truly fulfilling career awaits those who make the transition from academia to industry.

Chapter 9 Key Takeaways

* Academics is different from business but both are like businesses in that they rely on trust to continue to function.
* A business succeeds when it has customers and income above its operating costs.
* In industry jobs are expendable and the bottom line is the focus.
* Maintaining trust and customers are life in business.
* If you align yourself with the company's goals, you will achieve success in industry.

References

Cannon M. (2016). In the office: what's your role? HuffPost. www.huffpost.com/entry/in-the-office-whats-your_b_8001980 (accessed 11 June 2023).

Wong K. (2020). 9 tips for building trust in the workplace. Achievers. www.achievers.com/blog/building-trust-workplace/ (accessed 11 June 2023).

10

Choosing a Nonacademic Career

10.1 Dating Your Career

Dating and career exploration are similar. I know this may sound a little wonky, but in order to date effectively, you must first know yourself, then meet some people, and then go for it. You must have a clear notion of who you are and what you want to accomplish. Some criteria are important.

If you are at the end of your PhD and you are choosing to forgo the postdoc or are at the point of leaving a postdoc and are not pursuing academics, you are ready to start "dating" your next career move! As with dating, the career exploration process involves getting to know yourself, understanding yourself and your capabilities, and then discovering what opportunities are available, and ultimately going for them. It may be overwhelming to start, but you only need to take the first step. You must only make the next best move. You do not need to know exactly how, when, or in which direction in order to travel this entire path. Take one step at a time, get your bearings, and then take the next step.

Career Exploration (Dating) Process into Three Steps:

1) Getting to know yourself (strengths, values, talents, interests)
2) Dating some careers (what is out there, collect some data, make friends)
3) Putting yourself out there (resumes, referrals, interviews)

10.2 Getting to Know Yourself

So, let us start with some concepts. *Concept #1*: you are intelligent, and you know very well how to apply that intelligence to your goals in life. *Concept #2:* this will help you, but in this case, your intelligence will not help you as much as you might think in your career exploration process.

How to Make Your PhD Work: A Guide for Creating a Career in Science and Engineering, First Edition. Thomas R. Coughlin.
© 2024 John Wiley & Sons, Inc. Published 2024 by John Wiley & Sons, Inc.

Read these articles to help you understand that your intelligence could even be your greatest hindrance while you take this next step.

- Overthinking is killing you: science confirms you need to get out of your head (LaFata 2014),
- Smart people, dumb decisions (Hyman 2014),
- Smart folks are most susceptible to overanalyzing and overthinking (Belludi 2017),
- Science says this is what happens to you when you overthink everything (Morin 2017), and
- Six surprising downsides of being extremely intelligent (Lebowitz and Akhtar 2019).

If you do not think this is enough proof that being really smart here will not help you decide, then, go to google.com and type in the keywords: smart people decisions.

The second fact is – perfection does not exist.

To continue reading you must accept that you are handing over the reins. This journey is more like hopscotch, but in order to start you have to take that first "hop."

10.2.1 How Do I Know What I Want To Do?

Well, there are multiple ways to find that out. Do you find yourself in the middle of the room talking during parties or are you against the wall wondering how so many people have so much to say? Are you drawn to extroverts or introverts? It is pretty easy to figure out what you like more. Planned time with people and not large groups, or large groups and a bit of spontaneity. That is what they call introverts and extroverts and these are not too important, but we suggest assessing which way you lean.

In order to learn more about what you might find most fulfilling, we suggest that you take the Myers-Briggs Test to know your strengths and weaknesses. Here is the link to: www.16personalities.com (NERIS Analytics Limited 2011).

It is important to know what you enjoy doing. If you have worked in enough groups you may already know what you enjoy doing. Here are few questions to help: are you creative, analytical, detail-oriented, persuasive, good in front of audiences, a good communicator to multiple audiences, do you work well under pressure, do you like deadlines, do you like reading, do you like writing, do you like checking things for mistakes, do you like numbers, do you like motivating, would you rather get a bunch of work done for the group, be the face or presenter of the group, or do you like structure or free-form?

10.2.2 Foundational Questions

With the above figured out, you have laid a good foundation for the next step. Before you go on, we suggest answering the following prompts:

1 *What is motivating the next step?*

_____

_____

_____

2 *Is it stability? Working in a startup? Leaving the bench? Making money?*

_____

_____

_____

3 *What are the talents that I will bring?*

_____

_____

_____

4 *Do I want to work at the research bench?*

_____

_____

_____

5 *Work in more idealistic work like nonprofits?*

_____

_____

_____

6 *What do your colleagues and friends suggest?*

_____

_____

_____

Let us take a deeper dive into these questions. You can identify what motivates you and do your own discovery of careers with internships, coops, conferences, and informational interviews. Finding a career that meets your desires for your goals of either fulfillment, monetary, life balance, or some balance of these, provides a firm diving board to leap into the job application process. For example, if you like a comfortable lifestyle (and who does not), then it is important to obtain a job that enables comfort. In addition, if you want to be wealthy, it is important that you work toward applying for jobs that will get you there. Finally, if you are more motivated by your passions, then learning about the jobs that will fulfill you will be important for your success. Overall, employers appreciate when they meet a job candidate who is motivated, focused, and decisive in their career plans. Let us take some time to think about what motivates you so you can be firm as you approach different employers.

10.2.3 What Motivates You?

Consider fulfillment from work, career growth, pay, and work/life balance. Here are some example criteria to go into finding a job with:

1) **Balance**: some fulfillment from my job, steady career growth, interesting work, and reasonable work/life balance.
2) **Wealth**: fulfillment is not super important, pay and growth potential take precedence, and willing to set aside work/life balance.
3) **Fulfillment from Work**: fulfillment from work is paramount, while work life is not important so far as the work is fulfilling, willing to take less pay because of this.

Now, it is time to write your own criteria for the next job you are aiming for. Circle the importance of each of the following criteria below, fulfillment from job, career growth potential, wealth accumulation, and work/life balance.

Criteria	Importance for Next Job		
Fulfillment	Low	Medium	High
Career growth	Low	Medium	High
Wealth	Low	Medium	High
Work/Life balance	Low	Medium	High

Using the above exercise, list in order the criteria by priority, with the high priority first, followed by the medium priority, and then the low priority. Now, you have your priorities for searching for a job.

Importance for Next Job	Criteria (List Criteria)
High	
Medium	
Low	

10.3 Dating Some Careers

This next step is low risk and high reward. Exploring careers is a low-risk endeavor.

To find out what the careers are like and to find out what jobs are available does not commit you to a path, it only shows you what the path looks like and where they might go.

One of the best tools to evaluate careers is My Individual Development Plan (myIDP 2003) on www.myidp.sciencecareers.org (Fuhrmann et al. 2011).

myIDP allows you to explore career possibilities using questionnaires that assess your transferable skills, interests, and values. Taking this test will rank careers that most align with your skills and interests.

Using the top five results from myIDP, answer these questions and create five career personas:

Job Analysis One. Job Title/Career Direction: _____

Does it meet my need for what is motivating my next step?

_____

_____

What type of work does the job do? (pharma, government, venture research, etc.)

_____

_____

My transferable skills that I would use?

_____

_____

Subjects in this area that are new to me?

_____

_____

Subjects I find interesting?

_____

_____

Informational gathering?

_____

_____

What is the day-to-day?

_____

_____

Do you work on teams?

_____

_____

Is it remote or office or lab-based?

_____

_____

Does it have a career trajectory? (yes/no)

_____

_____

What are the hours?

_____

_____

Most appealing part about the job?

_____

_____

Least appealing part about the job?

_____

_____

Job Analysis Two. Job Title/Career Direction: _____

Does it meet my need for what is motivating my next step?

_____

_____

What type of work does the job do? (pharma, government, venture research, etc.)

_____

_____

My transferable skills that I would use?

_____

_____

Subjects in this area that are new to me?

_____

_____

Subjects I find interesting?

_____

_____

Informational gathering?

_____

_____

What is the day-to-day?

_____

_____

Do you work on teams?

_____

_____

Is it remote or office or lab-based?

_____

_____

Does it have a career trajectory? (yes/no)

_____

_____

What are the hours?

_____

_____

Most appealing part about the job?

_____

_____

Least appealing part about the job?

_____

_____

Job Analysis Three. Job Title/Career Direction: _____

Does it meet my need for what is motivating my next step?

_____

_____

What type of work does the job do? (pharma, government, venture research, etc.)

_____

_____

My transferable skills that I would use?

_____

_____

Subjects in this area that are new to me?

_____

_____

Subjects I find interesting?

_____

_____

Informational gathering?

_____

_____

What is the day-to-day?

_____

_____

Do you work on teams?

_____

_____

Is it remote or office or lab-based?

_____

_____

Does it have a career trajectory? (yes/no)

_____

_____

What are the hours?

_____

_____

Most appealing part about the job?

_____

_____

Least appealing part about the job?

_____

_____

Job Analysis Four. Job Title/Career Direction: _____

Does it meet my need for what is motivating my next step?

_____

_____

What type of work does the job do? (pharma, government, venture research, etc.)

_____

_____

My transferable skills that I would use?

_____

_____

Subjects in this area that are new to me?

_____

_____

Subjects I find interesting?

_____

_____

Informational gathering?

_____

_____

What is the day-to-day?

_____

_____

Do you work on teams?

_____

_____

Is it remote or office or lab-based?

_____

_____

Does it have a career trajectory? (yes/no)

_____

_____

What are the hours?

_____

_____

Most appealing part about the job?

_____

_____

Least appealing part about the job?

_____

_____

Job Analysis Five. Job Title/Career Direction: _____

Does it meet my need for what is motivating my next step?

_____

_____

What type of work does the job do? (pharma, government, venture research, etc.)

_____

_____

My transferable skills that I would use?

_____

_____

Subjects in this area that are new to me?

_____

_____

Subjects I find interesting?

_____

_____

Informational gathering?

_____

_____

What is the day-to-day?

_____

_____

Do you work on teams?

_____

_____

Is it remote or office or lab-based?

_____

_____

Does it have a career trajectory? (yes/no)

_____

_____

What are the hours?

_____

_____

Most appealing part about the job?

_____

_____

Least appealing part about the job?

_____

_____

10.3.1 Obtaining Informational Interviews

With these questions in mind, it is time to schedule some informational interviews. In order to obtain some informational interviews, you will need to use your network of people you know.

There are multiple people/networks that can be of help with this.

- LinkedIn.com
- Your undergraduate career centers
- Your doctoral career centers
- Your doctoral school may have a career networking system

It may take multiple emails to get an informational interview. This article recommends using an email template: "Introducing: The Email Template That'll Get You a Meeting With Anyone You Ask" (Frost 2020). The article may be true but still expect some failures; however, this article has the right format for the questions.

10.3.2 Informational Interview Questions

We recommend seeking out help in how to conduct an informational interview. There are many questions that can be used to assess (LiveCareer Staff Writer n.d.). Here is a list of questions that can be used to gather information through informational interview questions.

Must-have questions include:

- What is day-to-day life like?
- When do things get busy?
- What are the skills that you use from your scientific training?
- Do you work on a team?

- Do you work with people from multiple backgrounds?
- What is your career goal?
- Can we stay in touch, by connecting on LinkedIn (if you have not done so already), and sharing emails?

This article, "3 Steps to a Perfect Informational Interview," has some advice for getting ready for an interview (Zhang 2020).

Information gathering takes time; therefore, it is important to give yourself enough time to do this effectively. The time-limiting step is the informational interviews.

Conducting informational interviews with at least two people at distinct places per career path is sufficient. Interviewing people who are relatively new to the field as well as people who are in senior roles is very valuable.

10.3.3 Utilizing Your Institution

It is important to check out what is actually out there because no one will do it for you. So use this time to actually check out some of the programs that are available within your university or institution. In some cases, neighboring institutions will be able to help as well. So broaden your search and consider programs run internal and external to your institution.

10.3.4 Utilizing Networking Events

The networking events are vital to your success. They will expose you to multiple career paths and allow you to meet people who may have already transitioned out of the PhD and into a career path.

To find networking groups, you can utilize MeetUp.com, or simply Google groups in your area.

10.3.5 Having a Foundation

Once you have your informational interviews finished, you can compare the answers against your *Foundational Questions*.

10.4 Putting Yourself Out There

After you have narrowed down your career prospects, it is time to apply. Similarly, after you have done the informational interviews, you are in a much better position to apply.

At this point, you will need to be close to the end of your PhD or in a postdoc, ready to move on. Timing of this can be rather tricky, but it is of course doable, now that you have pointed your feet in a direction you want to go in.

Once you have that idea, start researching the companies in the area you want to transition into. Identify your geographical concerns, and begin contacting people at these companies over LinkedIn. Simultaneously, you will want to start working on your resume, detailed in Chapter 12.

Therefore, it is better to work through connections and referrals than it is to work just "spray" your resume and "pray" it gets noticed. Using your actual connections is more fruitful.

10.5 Conclusion

From this chapter, you learned how to seek information to learn what nonacademic career might be right for you. You learned how to map out your nonacademic career path. Through these steps, you will move more intentionally through your PhD or postdoc into mapping out your next step. In the next chapter, you will learn how to begin the transition out of academics by understanding the skills that are unique to PhDs like you. In addition, you will also learn how PhDs are ideal for remote work in this evolving and changing workforce and economy.

Chapter 10 Key Takeaways

---

* Start using LinkedIn immediately in your career to build your online network.
* Everyone has accepted LinkedIn as a way to reconnect with people from all stages of life.
* Conducting informational interviews with people in multiple careers can help you learn key insights into what careers are like.
* Learning from people in the workforce can help you rule in or rule out certain careers.

References

Belludi, Nagesh. (2017). Smart folks are most susceptible to overanalyzing and overthinking. Right Attitudes. http://www.rightattitudes.com/2017/08/30/the-dangers-of-overthinking/ (accessed 15 May 2023).

Frost, Aja. (2020). The networking email template that gets answers. The Muse. https://www.themuse.com/advice/introducing-the-email-template-thatll-get-you-a-meeting-with-anyone-you-ask (accessed 27 April 2023).

Fuhrmann, CN, Hobin JA, Lindstaedt B, and Clifford PS. (2011). Home page. https://myidp.sciencecareers.org/ (accessed 15 May 2023).

Hyman, Ira E. (2014). Smart people, dumb decisions. *Psychology Today* (21 August), https://www.psychologytoday.com/us/blog/mental-mishaps/201408/smart-people-dumb-decisions (accessed 15 May 2023).

LaFata, Alexia. (2014). Overthinking is killing you: science confirms you need to get out of your head. *Elite Daily* (4 September), https://www.elitedaily.com/life/science-overthinking/736245 (accessed 15 May 2023).

Lebowitz, Shana, and Allana Akhtar. (2019). Surprising downsides of being extremely intelligent. *Business Insider* (15 July), https://www.businessinsider.com/downsides-of-being-smart-2016-7 (accessed 15 May 2023).

LiveCareer Staff Writer. (n.d.). Questions to ask at the informational interview. *LiveCareer*. https://www.livecareer.com/resources/interviews/questions/information-interview (accessed 27 April 2023).

Morin, Amy. (2017). Science says this is what happens to you when you overthink everything. *Inc. Magazine* (25 April), https://www.inc.com/amy-morin/science-says-this-is-what-happens-when-you-overthink-things.html (accessed 15 May 2023).

NERIS Analytics Limited. (2011). 16Personalities: free personality test, type descriptions, relationship and career advice. http://www.16personalities.com (accessed 15 May 2023).

Zhang, Lily. (2020). 3 steps to a perfect informational interview. The Muse. http://www.themuse.com/advice/3-steps-to-a-perfect-informational-interview (accessed 15 May 2023).

11

Transitioning Out of Academics

11.1   How Do You Actually Convert Your PhD Into the Job You Want?

Whether you are in a postdoc, at the end, middle, or beginning of your PhD, there are methods to make your experiences appealing to potential employers.

A PhD typically possesses a variety of characteristics and traits that make them valuable to companies, including all of the knowledge and expertise they acquired throughout their years of education. Perhaps more significantly to employers, because the primary objective of a PhD is to foster an environment that produces an autonomous researcher, PhDs are frequently ambitious, independent, and highly resourceful individuals. These attributes that make them extremely desirable to companies.

11.2   Ideal for Remote Work

Not only are these qualities of independence desirable in the workplace, but they are also well-suited to the remote work environment. During the COVID-19 outbreak, remote work increased rapidly and exponentially. During this time, the "Stay at home" policy was promoted to control and mitigate the pandemic, while lessening the burden on national healthcare systems and economies (Anderson et al. 2022). Working from home, also known as remote work, telework, or mobile work, was promoted in industry and academic settings. From February to May 2020, the remote workforce increased from 8.2% to 35.2% in the United States (Saltiel 2020). Many businesses and educational institutions went completely remote when the pandemic began, and as many as 30% of all companies continue to employ most or even all remote workers.

How to Make Your PhD Work: A Guide for Creating a Career in Science and Engineering,
First Edition. Thomas R. Coughlin.

The attributes of independence, ambition, and resourcefulness that most PhDs possess are incredibly beneficial in the remote workforce. Companies are aware of this and prefer individuals who can manage themselves and work autonomously. PhDs are the ideal candidates for this work.

11.3 Transferable Skills

In addition to a PhD having the intrinsic qualities that make them great candidates for industry, they also possess transferable skills.

A transferable skill is a task, or a group of tasks or activities, that are applicable to academics and industry. In academics, PhDs master multiple skills that can be translated to multiple industry roles. Here is a list of some of the transferable skills that PhDs usually gain:

Analysis and Problem-Solving:

- Define a problem and identify possible causes
- Comprehend large amounts of information
- Form and defend independent conclusions
- Design an experiment, plan, or model that defines a problem, tests potential resolutions, and implements a solution

Interpersonal and Leadership Skills:

- Facilitate group discussions or conduct meetings
- Motivate others to complete projects (group or individual)
- Respond appropriately to positive or negative feedback
- Effectively mentor subordinates and/or peers
- Collaborate on projects
- Teach skills or concepts to others
- Navigate complex bureaucratic environments

11.3.1 Project Management and Organization

- Manage a project or projects from beginning to end
- Identify goals and/or tasks to be accomplished and a realistic timeline for completion
- Prioritize tasks while anticipating potential problems
- Maintain flexibility in the face of changing circumstances

11.3.2 Research and Information Management

- Identify sources of information applicable to a given problem
- Understand and synthesize large quantities of data

- Design and analyze surveys
- Develop organizing principles to effectively sort and evaluate data

11.3.3   Self-Management and Work Habits

- Work effectively under pressure and to meet deadlines
- Comprehend new material and subject matter quickly
- Work effectively with limited supervision

11.3.4   Written and Oral Communication

- Prepare concise and logically written materials
- Organize and communicate ideas effectively in oral presentations to small and large groups
- Write at all levels – brief abstract to book-length manuscript
- Debate issues in a collegial manner and participate in group discussions
- Use logical argument to persuade others
- Explain complex or difficult concepts in basic terms and language
- Write effective grant proposals

11.4   Favorite Skills

Having trouble deciding what nonacademic career might be right for you? Do not worry. You would not be the first PhD. The beauty of being where you are is that many PhDs have been there too. Looking at the list of PhD skills above, can you list, circle, or highlight your favorite skills? Then, look at the next table, which includes some of the skills that are prevalent in each of the jobs below (Table 11.1).

11.5   Matching the Skills

Did you find skills that you enjoy from your PhD in these positions? Not surprisingly many jobs available to PhDs require similar skills. There is a lot of overlap, but certain jobs will require not only your PhD expertise but also some additional talents, like writing, teaching, inspiring, discerning, selling, convincing, debating, listening, and much more. Take your time as you understand these jobs, but when you find one you like, do not hesitate. Go through the exercises presented in Chapter 9 to learn about the careers that you care about.

Table 11.1 Using the industry jobs identified in Chapter 8, the transferable skills that a PhD might need in a multitude of jobs are listed below.

Field	Job	Transferable Skills
Journal editing	Journal Editor or Senior Editor	• Reading, researching, writing, and editing papers and articles • Collaboration and teamwork • Critical thinking • Project management • Data interpretation • Field expertise • Researching • Defending independent solutions and decisions
Technology transfer and commercialization	Intellectual Property Liaison	• Researching • Field expertise • Communication • Project management • Critical thinking • Data interpretation • Teamwork • Leadership
	Innovation and Commercialization Manager	• Defending independent solutions and decisions • Learning new information
Startups	Startup Founding Teams	• Researching • Field expertise • Communication • Critical thinking • Project management • Data interpretation • Teamwork • Leadership • Defend independent solutions and decisions • Learning new information
Business management	Business Development Manager	• Researching • Communication • Project management • Critical thinking • Data interpretation • Teamwork • Defending independent solutions and decisions
	Management Consulting	• Learning new information • Writing reports

(Continued)

Table 11.1 (Continued)

Field	Job	Transferable Skills
Venture capital	Venture Capital Consultant	• Researching • Communication • Project management • Critical thinking • Data interpretation • Teamwork • Defending independent solutions and decisions • Learning new information • Writing reports
Marketing and sales	Medical Science Liaison	• Collaboration • Reading papers • Researching • Area expertise • Critical thinking • Communication • Personal management • Data interpretation • Learning new information • Writing reports • Relationship building
	Product Sales Specialist	• Reading papers • Communication • Research • Critical thinking • Project management • Area expertise • Personal management • Data interpretation • Learning new information • Relationship building
	Marketing Research Analyst	• Reading papers • Writing reports • Communication • Research • Critical thinking • Data interpretation • Project management • Area expertise • Defending independent solutions and decisions • Learning new information

Table 11.1   (Continued)

Field	Job	Transferable Skills
	Technical, Scientific, or Medical Writer	• Reading papers • Writing and editing reports and other print and digital deliverables • Communication • Research • Critical thinking • Data interpretation • Project management • Area expertise • Defending independent solutions and decisions • Learning new information
Product	Principal Engineer	• Reading papers • Writing papers and reports • Area expertise • Critical thinking • Applying engineering principles • Research • Communication • Critical thinking • Data interpretation • Data analysis • Project management • Defending independent solutions and decisions • Learning new information
	Scientist	• Reading papers • Writing papers and reports • Area expertise • Critical thinking • Applying scientific principles • Project planning • Research • Communication • Data interpretation • Data analysis • Project management • Defending independent solutions and decisions • Learning new information

(Continued)

Table 11.1 (Continued)

Field	Job	Transferable Skills
Regulatory	Regulatory Affairs Specialist	• Reading papers • Writing reports • Area expertise • Critical thinking • Applying scientific or engineering principles • Research • Communication • Data interpretation • Data analysis • Project management • Defending independent solutions and decisions • Learning new information
Equity research	Equity Research Analyst	• Reading papers • Writing reports • Area expertise • Critical thinking • Applying scientific or engineering principles • Research • Communication • Data interpretation • Data analysis • Project management • Defend independent solutions and decisions • Learning new information

11.6 Conclusion

As you learned from this chapter, transferable skills are skills that you can use in your current role and in a future job. While working in teams, collaborating on teams, or analyzing data from experiments, PhDs have an endless source of skills that can be applied to a multitude of roles. In the next chapter, you will see how these transferable skills can be incorporated into your resume.

Chapter 11 Key Takeaways

- ❖ A PhD transferable skill is a task or a group of tasks or activities performed in academics that are directly applicable in an industry setting.
- ❖ A PhD accumulates many transferable skills during their PhD that translate to many PhD-level jobs, outlined in Chapter 8.

References

Anderson, Roy M., Hans Heesterbeek, Don Klinkenberg, and T. Déirdre Hollingsworth "How will country-based mitigation measures influence the course of the COVID-19 epidemic?" *The Lancet*, 2022, https://www.thelancet.com/journals/lancet/article/PIIS0140-6736(20)30567-5/fulltext (accessed 15 May 2023).

Saltiel, F. (2020). Who can work from home in developing countries? Semantic Scholar. https://www.semanticscholar.org/paper/Who-Can-Work-From-Home-in-Developing-Countries-Saltiel/02b37ed279c7453d2415373f4a6704f23f95b177 (accessed 15 May 2023).

12

The PhD Resume

12.1 Prepare for the PhD Career Early

Waiting until the last months of a PhD to look for a job or career, leaves many PhDs scrambling to create a competitive resume. Ultimately, these PhDs can have a rocky transition and may spend time being unemployed. Even if your goal is to be a professor, it is best to start building a resume over the course of your PhD.

To build a great resume, it is important to ask yourself, "what do employers want?"

Another way to think of your resume is that you are showing someone why they should pick you to work on a team to carry out their vision. A PhD has the capacity to move into leadership roles in industry, where vision becomes more important, but starting out in industry, you will want to show you have the skills to do the job they are hiring for.

So in order to start crafting a resume, aim to have a rough idea of a career you may be interested in after your PhD. This career needs to be outside of academia (because in academics, resumes are not typically important).

12.2 Differences Between a Resume and CV

A resume and a curriculum vitae (CV) are two types of documents used in job applications, but they have different purposes and formats. Here are the key differences between the two:

A resume is a one- to two-page document that highlights a candidate's qualifications and work experience. It is typically used for job applications in the United States and Canada. A CV, on the other hand, is a more comprehensive document that provides an overview of a candidate's entire academic and professional

How to Make Your PhD Work: A Guide for Creating a Career in Science and Engineering,
First Edition. Thomas R. Coughlin.
© 2024 John Wiley & Sons, Inc. Published 2024 by John Wiley & Sons, Inc.

history. It is used primarily for academic, research, and medical positions, and is used more frequently in Europe, Asia, and Africa.

As mentioned, a resume is usually one to two pages in length, while a CV can be several pages long. A CV typically includes a detailed list of a candidate's academic and professional accomplishments, including publications, awards, presentations, and teaching experience.

Resumes are typically targeted to specific jobs and employers, while CVs are aimed at a broader academic and professional audience. A CV is meant to showcase a candidate's entire academic and professional history, rather than just their qualifications for a particular job.

In summary, a resume is a concise, targeted document that highlights a candidate's relevant skills and experience for a specific job. A CV is a more comprehensive document that provides a detailed overview of a candidate's entire academic and professional history. Understanding the differences between these two documents is important when applying for jobs in different countries or industries.

12.3   Begin With the End in Mind

Since it is best to begin with the end in mind, this chapter will review what a strong resume looks like and give you a format to use in your job applications. This format has been tweaked using many resources and publications on PhD level resumes and has worked for many job seekers that have utilized PhD source's resources.

There are many new ideas about formatting resumes and people are experimenting with Photoshop and Illustrator to combine neat ways of organizing the resume. However, a traditional approach using Word or Google Docs works well with job boards. For PhDs, it is more important to show the content and breadth of relevant experiences rather than focusing on the artistic appeal of the resume.

12.4   Six Seconds of Resume Time

The amount of time hiring groups spend looking at your resume can vary depending on various factors, such as the number of applications received, the level of interest in the position, and the size of the hiring team. However, research suggests that on average, hiring groups spend about six seconds scanning a resume before deciding whether to consider the candidate further or not.

Submitting an effective and impactful resume is a critical component of the job search process. To make sure your resume quickly grabs the attention of the hiring group, it is important that you follow some essential guidelines. This includes using:

1) A professional-looking but easy-to-read format,
2) Including relevant keywords throughout the document,

3) Emphasizing your most relevant experience to match the job requirement or position you are applying for,
4) Quantifying your accomplishments with proper measurement, and
5) Ensuring that no errors have been made after proofreading and editing.

By following these tips and tricks, you may be able to increase your chances of making a positive impression in those all-important first few seconds when someone reads your resume.

12.4.1 Key Words and Clear Formatting

Although it is fairly common knowledge now, most companies use programs that match keywords from your resume to a job listing. This technique can keep your resume from getting noticed even if you meet the qualifications. Before getting into how to write a PhD level resume, it is important that you understand some basic resume writing rules. Here are the important rules:

12.4.2 More White Space Is a Good Thing

Contrary to popular belief, filling your resume with every little detail you can think of about your experiences, is not a good idea. It is overwhelming to read. Instead, you can direct the reader to key parts of your resume by having a clear hierarchy of text separated by white spaces. This way, the reader can easily follow your career and be guided to the key parts.

12.4.3 A Resume Must Have the Key Words From a Job Posting

What does this mean? It means every job posting you submit a resume to should be tailored to that job posting. A reviewer who sees the same key words in the resume as the job posting will view you as a great fit. This way, you give yourself the best chance to succeed. Similarly, if the first step at the company is to use an automatic reviewer, the resume must be tailored in order to get past that step. No one wants to submit a resume that gets automatically rejected by a computer because it does not fit the requirements.

12.5 PhD Level Resume Template

Below is a resume format that will act as a template for you to document your skills and achievements throughout your degree. The resume template in Figure 12.1 has been proven to work, helping many PhDs get jobs in industry (Figure 12.1).

NAME

⚲ Address ⏱ phone number ✉ email in linkedin url

Key Summary

- Creative and analytical with experience in both subject of study and secondary area of expertise.
- Fast learner and hard worker, with expertise in subject.
- Recipient of $X in funding, with X publications and X international conference presentations.

Industry Experience

Collaborator, Company or University, State   Month/Year – Month/Year
- Past tense verb, leading to some result
- Past tense verb, resulting in xx findings

Focus and Research

NIH Postdoctoral Fellow, Company or University, State   Month/Year – Month/Year
- Past tense verb, leading to some result
- Past tense verb, resulting in xx findings

Research Assistant, Company or University, State   Month/Year – Month/Year
- Past tense verb and action, saving the lab xx in spending and obtaining funding
- Managed X researchers, fostering collaborative environment

Fellow, Company or University, State   Month/Year – Month/Year
- Directed funded project, leading to X number of cutting-edge publications
- Collaborated on fast-paced, cross-functional team and did this superlative

Awards and Classes

Award Name, Company or University, State   Month/Year – Month/Year
Course Training, Company or University, State   Month/Year – Month/Year

Extracurricular

Volunteer, Company or University, State   Month/Year – Month/Year
- Directed funded project, working with multidisciplinary team to achieve XX goal
- Collaborated on fast-paced, cross-functional team and did this superlative
Hobby that you have
- This is something to keep things light and show you are human

Education

Ph.D., Major, College, City, State   *Anticipated* **2015**
B.S., Major, College, City, State   **2010**

Figure 12.1 This is a template for a PhD level resume. This resume includes previous example experiences. At the beginning of the resume, three bullets encapsulate direct metrics that the PhD can use to demonstrate transferable skills and experiences.

12.6 Parts of the PhD Level Resume

The PhD level resume can be broken into a few key components: the key summary, industry experience (if you have it), academic experience, awards and courses, extracurriculars, and education. This chapter will explore what goes into each of these key sections.

12.6.1 Key Summary

The "Key Summary" area of your resume is a place to put three key bullet points that summarize your strengths for the job you are applying to. Even if you do not know this job, it is good to start putting superlatives into your resume. In the last bullet of your "Key Summary" it is important to list some outcomes or results of your PhD that make your PhD stand out. These metrics or numbers will be relatable to both academic and industry professionals. In industry key performance indicators (KPIs) are important and by showing that you understand that money leads to outcomes, you can put yourself ahead of the other applicants.

12.6.2 Industry Experience

The "Industry Experience" section of your resume will list any experience that you have in industry, including any collaborations, internships, or co-op experiences that you had. Listing a collaboration that your research group may have had and a role will give you an edge when building your case for an industry role.

12.6.3 Academic Experience

In the "Academic Research Experience" or "Academic Experience" section of your resume, you will want to list the major role that you had as Research Assistant (this is what your role in the lab that you were in is called) and any other fellowships or roles that you had. Furthermore, this is also where you would list your postdoctoral experience. The title of this section can vary, and depending on the role that you are applying to you can tweak the title. For example, if you are applying to a job where having an area of expertise is important, like medical writing or biotech equity research, then putting in "Cancer Research," for example, will be beneficial to showing you have knowledge of a specific sector.

12.6.4 Awards and Courses

The section, "Awards and Courses," in your resume is an area for you to demonstrate any particular training that may be applicable to the job you are applying to. Since at the time of building your resume, you may not know what job you want, this is a good place to start storing experience that you have had. In addition, it is good to start thinking about what courses or certificates you could strive for that may make you more competitive as you identify the job you want.

12.6.5 Extracurriculars

In the "Extracurriculars" section of your resume, you may add in areas of interest or volunteer interests that you have outside of your job as a PhD or postdoc.

With this section, it will help show a potential hiring board more about you. This can help differentiate you as a more human candidate than a cyborg STEM whiz.

12.6.6   Education

In this section, "Education," it is completely acceptable to list when you anticipate finishing your PhD. Furthermore, if you have already finished it, then that is great. Of course, add in your Bachelor's training as well. It is not necessary to add in a postdoctoral training here – this belongs in your "Academic Experience" section.

12.7   Writing Bullet Points for Your Resume

When seeking how to write the best bullet point for your resume, there are a number of resources. Culminating all of the literature available on resumes, it is best to use action verbs that describe a role that you had followed by the result that it led to (Haseltine 2012).

When you are ready to apply to jobs, it is important to tailor the action verbs to the action verbs that are inside the job description you are applying to. This way, you will seem even more like the perfect candidate and give yourself the best opportunity that you deserve. This is also true of skills that are listed in the resume. The job description will list skills or training that is needed, and to qualify for that job you must list these same skills in your resume.

For example, in a mechanical engineering position, they may list that a primary qualification is that the applicant has experience in Computer-Aided Design (CAD) modeling. As such, it is important you list "CAD modeling" in a bullet point in your resume. Perhaps you may write, "Collaborated on CAD fluid dynamics model of Boeing 737 wing tunnel, leading to reduction in excess material."

12.8   Applying to Jobs Through Your Network

As you go through informational interviews detailed in Chapter 9, it is important to stay in touch.

Networking is a critical component of the job search process, and it plays a significant role in helping candidates find new employment opportunities. Building a strong professional network can open up doors to new job opportunities, provide valuable insights about the job market, and connect job seekers with influential individuals in their field.

Here are some of the key reasons why networking is so important when applying for a job:

12.8.1 Access To the Hidden Job Market

A significant portion of job opportunities are not advertised online or in traditional job postings. These jobs are often referred to as the "hidden job market," and they are typically filled through referrals from individuals within a company or through professional networks. Building a strong network can help job seekers tap into these opportunities and increase their chances of finding a job that aligns with their skills and interests.

12.8.2 Insider Information

Networking provides access to valuable information about companies, industries, and job markets that may not be readily available through traditional job search methods. By connecting with professionals in their field, job seekers can gain insights into industry trends, job requirements, and potential employers' hiring practices, which can help them tailor their job search strategies accordingly.

12.8.3 Referrals and Recommendations

Employers are more likely to hire candidates who come recommended by someone they know and trust. Networking can help job seekers build relationships with individuals who work in their target industry or company, increasing their chances of receiving a referral or recommendation for a job opportunity.

12.8.4 Professional Development

Networking can provide job seekers with access to mentors, coaches, and other professionals who can offer guidance and support as they navigate their careers. These connections can be invaluable as job seekers looking to improve their skills, gain experience, and advance their careers (Figure 12.2).

12.9 Conclusion

Overall, networking is a critical component of the job search process, and it can significantly increase job seekers' chances of finding new employment opportunities. By building a strong network, job seekers can tap into the hidden job market, gain valuable insights about their industry, receive referrals and recommendations, and access professional development opportunities.

Figure 12.2 After much research, planning, retooling your resume, and applying, you too can join the many PhDs in science and engineering leading satisfying careers in nonacademics.

Your resume is a fluid document. The sooner you start working on it and updating it the better it will be. It is not uncommon to have multiple resumes saved on your computer for all of the jobs you are looking for. In fact, you should have one for each job you apply to. As mentioned in Chapter 10, looking for jobs is like dating, if you do not present yourself properly, you will not be able to get the job. Take the time to tailor each resume.

Chapter 12 Key Takeaways

- ❖ A resume is not a CV. A resume consolidates your experiences and is tailored so that the resume is most relevant to a particular job posting.
- ❖ Having white space on your resume is not always a negative, and instead can help the human resources department look at your resume for key parts.
- ❖ The resume should have a hierarchy with key points up front so the reader can easily see why you are relevant without reading every line on the resume.

Reference

Haseltine, D. (2012). Job-search basics: how to convert a CV into a resume. *Nature Immunology* 14 (1): 6–9.

Transition Story: Leon "Jun" Tang, PhD

PhD: Icahn School of Medicine, PhD
Field of study: Biomedical Sciences
First position out of PhD: postdoc at Memorial Sloan Kettering Cancer Center
Current career position: Founding partner at InScienceWeTrust BioAdvisory

Undergraduate education: In 2001, when I pursued my undergraduate degree at Tianjin University, biomedical engineering was an emerging field that promised future economic rewards and job security. By the end of my undergraduate degree, the promise of these jobs was not there yet. There were not enough biotech jobs to support the oversupply of newly minted biomedical engineering undergrads.

Motivation for pursuing a PhD: I had mixed incentives when I chose to go for a PhD. My primary concern was creating opportunities for myself in the future. I was born in China and come from a modest and humble family background and upbringing. I chose my path based on securing a future career. For this reason, I chose to pursue a Master's degree and was admitted to Nankai University.

Master's Degree: I was careful about choosing my major. Instead of pursuing a passion, I looked ahead to find a major that would secure a future career. In 2008, I received my Master of Science degree in Biochemistry and Molecular Biology from Nankai University. I graduated with honors and published two papers, and the field of biochemistry and molecular biology afforded me the opportunity to transition into a PhD position in the US.

Start of PhD: I started my PhD at Icahn School of Medicine at Mount Sinai in Manhattan, New York. I was paid a stipend of approximately thirty thousand dollars per year and this was roughly more than many of the socioeconomic groups in China. I was pretty satisfied with my career move up until that point.

Environment: Driven by the desire to foster a better life for myself, I worked hard. I was not motivated by the idea of being a professor. Instead, I wanted to make a better situation for myself. The resources to be successful were readily available in my professor's lab. The lab was funded and there was enough money available to do work.

Project: I found biology to be a great research field to work in. Perseverance and wisdom are rewarded fairly. For me the paradigm of hard work equaling results was quite rewarding.

Advisor: After showing a strong work ethic, I was able to cultivate a strong relationship with my advisor. Our relationship became one of mutual respect. My advisor was honest with me and pragmatic. I really appreciated that about his approach. He taught me how to be successful in academics and that 'papers' are the key performance indicators to academic success.

Of course, the transition to the United States was not without challenges. The difference in language and slight challenges in culture created some hurdles for me to overcome. New York City is a tough place for a non-native speaker and navigating it was tough. I had to adapt to cultural differences and workplace dynamics.

Getting the next position: At the end of my PhD, it was very challenging to get a job in academics. I did not have the proper Visa status. That was a major reason for me to do a postdoc. I had to fix my immigration status. That is a huge hurdle in the system and many companies and programs have a lot of leverage with international students who desire citizenship or security within a foreign country.

Moving to a position at a nonprofit: At the end of my postdoc, I chose to pursue a position outside of academics. I took a detour and got into a venture fund in oncology and it overlapped with my PhD and postdoc study. I learned a lot in this new position and also completed level #1 of my chartered financial analyst (CFA) exams. After three years at this nonprofit, I learned enough from the nonprofit, that I decided to switch to something new.

Moving to the financial sector: After the nonprofit position, I moved into the role of Assistant Vice President at Barclays Investment Bank. In this position, I worked with and managed a team of five top analysts. I left Barclays and joined a company and became a Contractor in business development focusing on guiding life science clients' portfolio strategy and commercial planning decisions at the intersection of science and business. From this position, I decided to start my own company, where I currently work. In this position, I focus on valuing and analyzing new emerging biotech in China.

For someone considering a PhD who lives in China: In today's market, there is opportunity for scientific advancement in China and there has been a lot of investment into academic research, bolstering the status of many of the institutions to a high tier. Because of this, I would not suggest that PhDs come to the United States unless they desire to stay and live there. There are such differences in the workplace cultures between the United States and China. I would suggest if a Chinese student wanted to pursue a career in research in China, then they should complete the PhD in China. By contrast, if they wanted to come to the United States, I would recommend conducting a PhD at a top-tier university, like an Ivy League institution.

In addition, many PhD students who are willing to travel open up a vast world of opportunities to themselves. There are many great institutions in other countries and in order to be successful in an academic and research perspective, the course of study is more important than prestige. If the university has prestige, but you are unproductive, your degree will be hollow and this will be obvious to employers and you will have a steep hill to climb. So I recommend a challenging PhD. It is important to truly spend these years of your PhD learning to problem solve. I have found these skills to be essential for my later career in the financial sector.

To a PhD student who is not in a good program: With respect to absent advisors or a lack of funding, I would suggest that a PhD student simply move on from a poor situation. There is just too much economic cost to pursuing a PhD. Having a bad PhD just incurs too much economic cost. In other words, a PhD is too time intensive with too much implicit cost to have it be full of a lack of training or wasted time.

Looking back: The PhD was not a waste for me. The PhD fundamentally trained me how to think. The PhD helped me identify a problem, not only for my research but also for my life.

Advice for getting jobs: When considering advice for getting jobs, your "people" network is very important. I would advise PhDs to ask questions and talk with friends, peers, colleagues, and former classmates in their network to gain perspective on their careers to gain a better understanding of what you can achieve.

Dr. Tang is the author of 33 journal publications with an h-index of 33, 4,470 article citations, and 9 nature publications (Google Scholar).

Key Insights:

- Your PhD is yours and you are never without opportunity. There is always the option to change things if they are not working.
- Unlike experiments, you are able to navigate jobs, not just by trial and error, but also by talking to people and learning from their experiences.

Transition Story: Elizabeth Agadi, PhD

PhD: Integrated Biomedical Sciences, University of Notre Dame
First position out of PhD: Graduate Career Services Consultant
Current career: Stay-at-home mother

Motivation for PhD: I conducted my undergraduate career at the University of Notre Dame, where I studied biochemistry. I started to consider going for a PhD after a successful summer of undergraduate research and then finishing the core molecular biology course. I found myself not wanting it to be the end of studying biology. I wanted to explore and understand more about how the body's systems worked. This yearning for more, drove my decision to aim for a PhD program. Truthfully, I did not have a career in mind after the PhD. I thought I would enjoy teaching. I had looked at jobs in pharma but thought that it was not the right fit for me.

For my PhD, I stayed at the University of Notre Dame in the integrated biomedical sciences program, a new interdisciplinary program. In this program, I rotated through four research labs in the first year of my PhD. Overall, the rotation experience in each of the four labs was solid. I picked an advisor where I enjoyed the research and thought it would be a good fit.

In the middle of my second year, my first PhD lab closed and I transitioned to a very prestigious lab with a well-known principal investigator (PI). This lab was a good and natural step for my PhD.

Advisor: My PI was an amazing advisor and overall just an amazing person. For instruction and advice, my PI met with each of her new PhD students once a week in a one-on-one fashion and had a full lab meeting about once a month. As I became more independent and more trust was established, I met with my PI less frequently and we communicated mostly over email in between meetings.

How to Make Your PhD Work: A Guide for Creating a Career in Science and Engineering, First Edition. Thomas R. Coughlin.

Environment: My research group was a supportive, positive, and collaborative environment. It was an all-around good environment. I have heard of labs that are toxic and the lab I joined was not at all a place where people were trying to one-up each other. There were many senior lab members who provided practical training and guidance.

Project: My PI likes to give one project that is a "low hanging fruit" experiment and then one more stretch project. The "low hanging fruit project" is meant to be a no-brainer that we can definitely publish, while the stretch project was meant to be more difficult and not as definitive. Since a PhD can be pretty challenging, having the "low hanging fruit" project was a huge comfort. Some advisors throw you in the deep end without support, which can be challenging for graduate students' mental health as well. Overall, my PhD was not like that. I published two first author papers and each one had a bunch of coauthors from the collaborative nature of the group.

Career exploration: My PI was very supportive of my growth. During my PhD, I went to Haiti to explore global health work. Then, I got to do some advocacy work in Washington DC. I also gave a guest lecture for a course in my field. I did not have to be a teaching assistant because I had grant funding. Overall, I started my PhD thinking about teaching or doing a competitive postdoc following the academic track to become a tenure track faculty. But as I saw my advisor, an R1 research professor, and the 24/7 nature of the lifestyle, it was not something I wanted. I did not think I would thrive doing that.

I realized after the fact when I gave talks on career exploration, I thought my career would come out and be obvious. However, I explored things haphazardly, and I was not asking the important questions "what are your career values and where would you thrive?"

Changing directions: About halfway through my PhD, I started to feel burnt out. My PI was supportive of me and gave me freedom to explore different career options. I learned about global health, public health, and these paths led me to apply for a fellowship with a nonprofit in my college town's area whose mission was regional talent retention and supporting local businesses and organizations.

I received this nonprofit fellowship, which would begin immediately after completing my PhD, and I asked my PI for help with the decision. I was concerned that if I took the fellowship I was cutting off my options to get back to academics. My PI assuaged my concerns by indicating that if I stayed outside of academics for a year at this fellowship while still publishing, it would not hurt my academic career and I could continue on to a postdoctoral fellowship. However, if I stayed for two years, it would be difficult to get back into the academic world.

I took the role in the nonprofit and I liked some of the work in higher education that I did and learned some of the work that I also did not like. From there, I applied for a position as a Graduate Career Consultant to Science PhDs at Notre Dame. I obtained it and then stayed at this position for four and half years. Honestly, during my PhD, student advising as a career had not occurred to me, but it ended up being a great fit and there was enough interest for me to keep staying in this work.

Throughout my PhD, my PI was very motivating and she even encouraged me when I was discerning the role of Graduate Career Consultant, sharing that she thought it would be a good fit for me.

Advice for a PhD student: My advice for PhDs is to be intentional with their career exploration and planning. I did not take the time to really dive into this but as a Graduate Career Consultant, I learned that it is important to take professional development seriously. The process can seem difficult but it is worth it to find what you might really enjoy. That said, it is hard to explore jobs when you are stressed and writing your dissertation, so start early! It is worth taking the time to reflect to know yourself better and to pursue a career that is aligned with your skills, interests, values, and personality.

Key Insight:

- Taking action toward your personal career development is important. When taken seriously, you are able to more purposefully obtain a career you desire. By contrast, not taking action, can lead you to have a narrow view of the jobs you can obtain, leaving you without many options.

Transition Story: Laura Zheng, PhD

PhD: Johns Hopkins University
Field of study: Neuroscience
First position out of PhD: Medical Writer
Current career position: Sr Data Scientist in health insurance

Motivation for pursuing a PhD: When deciding to begin my PhD, I had been working at my first job in an academic research laboratory. Suddenly, the 2008 financial crisis happened and I was searching for a job within a market impacted heavily by the recession. At this time, most people were just worried about getting a job or keeping the one they had. To me, it seemed like a good plan to go to graduate school for several years and wait out the bad economy, and then emerge back with higher credentials when the economy was doing better. With this in mind, I chose to do a PhD in public health because I found this area of research very interesting. During my undergraduate degree, at Cornell University, I had thought that being a professor would be a fitting and good career for me. With the economic factors and some semblance of a goal of being a professor, I entered graduate school.

Advisor: I was accepted to go to Johns Hopkins Bloomberg School of Public Health for graduate school to get my PhD in Environmental Health. I was extraordinarily lucky that I had a really great PhD advisor and I think that was a huge factor in contributing to my success as a PhD. My advisor was always involved with my progress. In times when I was struggling, my advisor would provide one on one mentorship, where they taught me how to problem solve, write papers, and conduct analyses. I learned so much from my advisor, not just in terms of how to conduct research, but also how to mentor people.

Environment: During the PhD, I was in a fantastic group. We were all very supportive of each other. Additionally, we would reach out and help one another with different parts of projects. For example, when I tried to learn the R programming language for the first time, I was having a hard time making these very specific plots for my paper and I was constantly asking another student for help. My fellow group mate was very gracious and supportive even though I must have been quite bothersome. That really set the tone for me in my group. I really loved having that guidance early on, and I tried to pay forward the favor by making myself as available as possible to newer students. By doing this, I was able to continue cultivating that collaborative and helpful environment.

Project: Overall, my project was good. Working on it taught me many of the skills that I use today and I truly felt that I had accomplished something. However, that said, I was not super picky about the work I was doing and I was more open to new ideas and projects. I know there are some people who are very particular about their projects. For me, I was more agnostic toward my project. To me, all projects are interesting and an opportunity to learn something new.

Career support: At my institution, at that time, there was not much talk about career options outside of academia or the government, such as the Environmental Protection Agency (EPA) or the Food and Drug Agency (FDA). It was an unspoken expectation, at least in my department, that one would go into one of these career paths. For example, I remember a discussion I had with an alumnus who went into management consulting where he said that his advisor took it almost as a betrayal when he learned that his mentee's job was in private industry. For me, jobs in academics or government were the only jobs that I thought I could get with my PhD. It was not until I arrived in New York City for my postdoc that I learned about many other job options for people with PhDs.

Postdoc: In deciding what to do after my PhD, I had no idea what I wanted to do after I graduated but I was beginning to think that I should not stay in academia. However, at that time, I wanted to live in New York and chose to take a postdoc position in Mount Sinai in Manhattan, NY. Despite not wanting to enter academics, I thought that strategically, being in New York City would expose me to careers and job prospects in nonacademic paths. To my delight, my hypothesis for my career move worked.

In New York, I was exposed to many different careers. I attended networking events, like "What Can You Be With a PhD?" and learned about careers in both medical communications and data science and many other PhD careers. When I spoke to other PhDs around the country it appeared to me that there was more infrastructure in New York for transferring careers than other places.

First Industry Position: After my postdoc, the first stop was at a medical communications agency firm preparing slides for keynote speakers at major oncology conferences for pharmaceutical companies. I liked my coworkers and enjoyed learning how that entire ecosystem (e.g. pharma clients, oncologists, med comms agencies) worked. Although I was acquiring new skills in this role, I missed crunching data. When that ended, I took another chance and transitioned into a different field.

Career Transition: When I was a postdoc, several friends had transitioned into data science via the Insight Data Science Fellowship, an intensive 7-week postdoctoral training fellowship bridging the gap between academia and data science. I had a chance conversation with other alumni from Mount Sinai who suggested I apply to the program as well. After consulting with my friends, I put in my application, interviewed, was accepted, and then started the program. The program was incredibly helpful and allowed me to gain insights into what industry employers were looking for in the data scientists they hired. After months of research, applications, and countless rejections, I finally landed my first job as a Data Scientist in Credit Risk Modeling at a fintech startup. In this role, I used my combined data science and machine learning skills from my PhD, postdoc, and newly minted Insights Data Science Fellowship.

Second position and current role: Since this position, I transitioned into a new role at a major health insurance company. This company was recently acquired by a large national retail pharmacy, and I transferred from the retail pharmacy side of the company to a new role in the company, which is now my role in data science. I am pretty satisfied with the work that I do. I feel like I am fairly compensated, my coworkers are great, and finally, I see potential for advancement in my future career. I envision that I will stay here for a while.

Looking back: It has been a really long road getting here with a bunch of trial and error in between. It would have been really nice if I had arrived at my current career straight out of a PhD and I did not have to spend 3–4 years trying to find my way, but the past cannot be changed. I feel good where I am now because I have had the experiences of being in places that were not a good fit so I know what a good fit feels like for me and how important it is for my future.

Advice for future PhDs: The most important takeaways are that you might have to deal with a lot of rejection and that is a big wake-up call for a lot of PhDs. I spent my life being a really good student and I was used to always getting accepted, and then there is a situation where you are always getting rejected and that is such a big shock. You might be good at everything you do, but you are sometimes going to run into things that might not work for you.

In the work world, the one thing I might have wished I knew before getting a job is how collaborative work is in industry. I wish I knew how my role would work within a larger company earlier on in my career. That being said, once I got the hang of industry and how it worked, I quickly learned that I was specially trained to dig deeper into problems than most other individuals. With their scientific training, PhDs are trained to be able to look at data and understand and question the norms with a highly analytical and deep problem-solving. There is a learned patience and persistence that others might not have to go deeper into the problems. With that, PhDs are highly valued in industry settings.

Finally, historically, a PhD would go into a faculty position but that is not the case anymore. Universities are training more PhDs, but they are not increasing their faculty hiring, which means that there is more competition for those academic faculty positions. Naturally, this means that many PhDs will need to look elsewhere for their jobs. Another thing is that it can be tricky to get hired for your first PhD job because many employers are not sure if you know how to function in a work environment. Unlike some other professional programs (medicine or law) there is not a structured career trajectory that you follow anymore. It is more of a "choose your own adventure" type of experience, where there are different career paths that you can go into but you have to decide what you want to do and the path to get there can be less clear.

Key Insights:

- Although academics can be collaborative, industry relies on teamwork and communication with a collaborative approach.
- The value of a PhD is not just in their specialization and deep knowledge of a specific field, but also in the skills they developed, like problem-solving, data interpretation, self-management, research, and communicating complex ideas to diverse audiences.
- With a PhD the path is uneven and there is no direct solution for PhDs looking for jobs after their PhD.

Transition Story: Amar Parvate, PhD

PhD: Purdue University
First position out of PhD: postdoc at Scripps Institute
Current career position: Cryo-EM microscopist and Biochemist Staff Scientist at Pacific Northwest National Laboratory

Motivation for pursuing a PhD: I completed my Bachelor's and Master's degrees in classical biology while studying in India at Savitribai Phule Pune University. I considered conducting my PhD in India, but in India, open seats for PhDs in these positions were and still are limited. Instead of trying to earn a PhD in India, I wanted to go to a place with a "state of the art" education system. My research showed countries with high-caliber graduate education programs are located in the United States, Germany, Singapore, Japan, Australia, and Canada. However, none of these countries matched the scale of the programs offered in the United States of course, Germany has the Max Planck Institutes, but just in the state of California there are six or seven University of California (UC) institutes, and that is just one state. Therefore, I decided I would have more opportunities and exposure to a wider research community in the United States.

I applied to about eight or ten PhD programs in the United States. I was interviewed by four institutions and was offered two positions. I selected Purdue University for my PhD program. Part of the reason why I selected Purdue over my other choice for my PhD was because it has one of the single largest evolution, immunology, and biochemistry departments, and I did not have to declare one of the sub focuses as my area of study. I could declare my focus broadly and rotate through the labs. Being that I was living in India, I had not seen these labs in person, so the idea of rotating through more labs with more options was quite favorable. Plus, Purdue was well known for its engineers, such as Neil Armstrong.

During the first year of my PhD, I rotated through a classic microbiology lab, then in a lab focused on protein crystallography, and then an electron microscopy

How to Make Your PhD Work: A Guide for Creating a Career in Science and Engineering,
First Edition. Thomas R. Coughlin.
© 2024 John Wiley & Sons, Inc. Published 2024 by John Wiley & Sons, Inc.

(EM) lab. I liked EM and joined this research lab. Science-wise, it was an easy decision, and it also had a lot of opportunity for growth, because after I joined, the field has continued to expand.

Advisor: My PI was young, and was not tenured. Early on, it became evident that he did not have much experience training PhD students. To make matters worse, my PI was having issues in his personal life and was overall, facing burnout, which in the end, resulted in him not applying for tenure. In fact, he no longer works in STEM. Having a PI who lacked focus made the PhD tough.

Project: My project was in EM. All the projects in my PI's lab were very ambitious. So much so that the experiments were nearly impossible to complete. In fact, I still look at the literature and work in the field today, and even five to ten years later, the work he proposed still has not been done. Because of this, I had to completely rethink my project. I remember going to my PhD thesis committee and asking them if they cared if I changed the scope of the work. They said as long as I am able to do experiments and present something at the end of my PhD I would be able to graduate. This helped me gain some focus on starting reasonable experiments. My PI's projects were just too big and hard to understand at the time.

Environment: The working environment was heavily influenced by my PI's lack of focus. The lab was toxic for a long time and it was confusing because our PI was under so much stress in his own life. Also, the oversight on the granularity of the work made us all quite confused.

I remember being in this situation and thinking that it was not right, but it still took six to eight months to make a change. No one was looking out for this PI and no one was looking out for his graduate students. I was left to go through this situation and sort it out. That is the way it was.

In the end, myself and the other two graduate students and one postdoc student moved on. The three PhDs, including myself, had a difficult time finding labs to go into. After talking to a few PIs, I was given a second PhD PI and she was not a subject matter expert in EM, but she helped me find fellowships, and funding and read through my thesis. She was excellent and really wanted to help me.

Next steps: During this whole process, perspectives from my wife and mentors helped greatly. I was able to gain perspective on my personal situation. My logic was not to be a faculty member. I knew that even if I was competing with other postdocs for a faculty position that the competition would be incredibly difficult and I did not want to be a faculty member enough to face this level of competition. Industry careers were difficult to come by as well because I needed Visa sponsorship, so I decided to do a postdoc. My second PI helped me land a prestigious postdoc at Scripps Institute. When I got this postdoc appointment I was able to wrap up my thesis and graduate.

Postdoctoral position: Once in San Diego, CA for my postdoc at Scripps Institute, I worked with Erica Ollmann Saphire, PhD. She was an amazing PI. She was positive, and the environment was collaborative and engaging. I still only wanted to do a postdoc for two to three years and was very honest with her. She said as long as she knew what I wanted from the postdoc and was honest with her, then we would be productive and it would be easy for her to help me navigate the next steps. The whole environment was amazing and people were collaborative and the conversations with my peers helped accelerate research. It was the best experience of my life.

After one year in my postdoc, the COVID-19 pandemic started and I was glad to be in a secure postdoc position and my work was not too affected by it. Then, after two years in my postdoc, a national lab connected with me. I was under the impression that they only employed US citizens, but they helped me with my Visa and they had me come there and work.

Current Position: I transitioned to a position of Cryo-EM Microscopist and Biochemist Staff Scientist at Pacific Northwest National Laboratory. I am very satisfied with this position and overall the environment is very positive. It has been great to still do work inside my area of expertise from my PhD.

Advice for future PhDs: I would want a PhD students to know that they are not alone and there are always other options. A lot of people think there is no fallback situation and they are stuck, but that is not true. You can move and you can find situations and you can reach out for help. If you mentally think you are caged, then try to find help. Every university has resources for PhDs.

Looking back: In the end, I would do the same choices as I made because I ended up having amazing advisors. From the opposite side, being a faculty member does not come with a guide book and that is not easy, but I choose team graduate students. My advice is, if a toxic situation occurs, then do not hesitate to act and try to do what you need to make it better for yourself.

Key Insights:

- Despite challenging environments and changes to funding or your PIs tenureship, there is always the possibility to ask for help and seek out additional resources.
- The environment in the research lab is influenced by the PI. A PI with focus, drive, and excitement in the work is the best match for completing a PhD.

Transition Story: Henry Cham, PhD

PhD: Baylor University
Field of study: Neuroscience
First position out of PhD: Data Scientist in a startup
Current career position: Data Scientist in home property insurance

Motivation for pursuing a PhD: As a kid, I was always interested in science. I wanted to be a scientist. I wanted to ask my own research questions and work on them freely to explore knowledge. That was my goal.

I attended undergraduate school at SUNY Binghamton in New York, where I chose to major in psychology. While at college, I conducted some undergrad research in a psychology lab, and out of all the things that I did in college I really enjoyed research and the process of doing research. I loved talking with the other grad students, reading journal articles, and doing the experiments. It was really from that experience that made me want to pursue the PhD.

Starting the PhD: I decided to take a year off from undergraduate school and then apply to a few PhD programs. I interviewed at a few graduate schools and really liked Baylor University in Houston, Texas. Baylor had exciting work and a group of neuroscience professors who were published in prestigious journals. I knew at that time that funding was an important metric to look for in a PhD program, and these professors in the Neuroscientist Department at Baylor seemed to have enough money to fund graduate students. With the robust program, the strong group of advisors, and the presence of funding, I decided that Baylor was a good fit. On top of all that, I was excited to explore Houston, which was a fun city with a lot to offer.

PhD rotations: In my first year, I rotated through different neuroscientist professor's labs. Truthfully, I did not really fit into the labs I rotated through. I wanted to go into research related to systems neuroscience (a subdiscipline of neuroscience and systems biology that studies the structure and function of neural circuits and systems) but did not find a lab that shared my vision in my department. After all of the rotations through the various professor's labs, I decided to go into one of the last labs I rotated through because of a few positives: a small group, a motivated advisor with funding, and the research seemed exciting enough to keep my interest.

I was not prepared for the repercussions of the sudden twist that my PhD had in store for me. After a few months, the research advisor that I had chosen to mentor under let me know that I would be moving to his wife's laboratory to do research there. He would still be my thesis advisor but I would report to her as my research advisor and do her research.

Project: It was quite a change. The husband's lab was made up of 4 to 5 students and the wife's lab was made up of fifteen students with postdocs. At that time, my new research advisor was making the transition from doing primate research to conducting neurological research on mice. Because it was a new animal model, I had to learn how to do surgery on mice, learn electrophysiology, and how to perform readings on mice. My advisor was absent, and I realized that she did not have the skills to teach me data collection. Luckily, there were postdocs in the lab who offered some support, but I mostly had to learn the techniques I used on my own.

Advisor: Over the course of the first few months of the start of being in my unexpected research advisor's lab, I began to realize that she was neither available to help guide me nor her other PhD students in our research topics. Instead of regular meetings, which I thought was the norm, we had irregular meetings, getting together on average, every four weeks. My advisor mostly traveled to conferences, and I was left to find my own mentors through the postdocs. As a new PhD student and young scientist, I found this very challenging.

Environment: My research advisor demanded results. It seemed like her way of doing things was pushing her research group to create results without nurturing our development. She expected us to sort out our development ourselves. Ultimately, this approach cultivated an environment of high stress. Because of this, there was high turnover in the lab. And as such, the high turnover and lack of a consistent member of the lab did not allow for much knowledge transfer in the lab. There were not enough long-term researchers in the lab to be able to help with troubleshooting existing projects and technological issues. Unfortunately, in general, my advisor could not help with the experiments as she focused on writing papers, getting funding, and interpreting data.

A tough decision: Within six months of working in the new lab, I reached out to the head of the department about the difficulties I was facing in my research advisor's lab. I told him that I did not like working where I was and asked him if I should stay in the lab. He said that if you think your research is putting you on a path to graduate with your PhD, then you should continue. He told me that obtaining a PhD from a lab is a great achievement and if you think your research project will allow you to graduate from this lab, then you should stay. By contrast, he said, if it's impossible to publish and graduate, then you should switch labs.

At that time of my PhD, there was in fact another lab that I could transfer into. I liked speaking to this advisor when I interacted with him and his students seemed happy. I really considered moving to this lab, but overall, his lab focused on research that was not in my core interests. And with the advice I had just received from the member of the department's advice in my head about how important it is to have a project that leads to a publication and eventual graduation, I decided to stay in the lab I was in because I was making headway on the research.

Despite the lack of presence from my advisor, the project was going fairly well and I felt that I was progressing along.

Challenging times: The remaining four years of my PhD were very tough. There was a lot of anxiety. I was really unhappy for those years. There was a lot of chaos. I could not trust my professor to have my back. I was really burnt out for about 3 years. I was complaining all the time. And when I talked to my colleagues, they complained all the time. I even developed headaches from all the stress.

Preparing to graduate: Finally, in my fifth year, it was the time to plan my graduation, but I did not have my publication. My research advisor told me that she wanted me to work and stay for 2 more years, making it a 7-year PhD. I assumed that she saw me as a reliable person and thought that she could collect more data from me. The prospects of staying a few more years made me really unhappy and I wanted to leave as soon as possible.

On my own, I met with each of the members of my dissertation committee and told them what was going on in my lab and that I did not want to continue. I did not want to do another project during my PhD. With my individual conversations with my dissertation committee and their support of me finishing my PhD, I met with my research advisor and committee. In this meeting, I announced that I did not want to do another project. My research advisor was surprised by my adamant nature and she began to push back and complain a lot. Luckily, I was officially still assigned to her husband as my thesis advisor, and therefore, I did not need her permission to graduate. That said, I did need a publication.

After this landmark meeting, I told my research advisor that I wanted to publish the data I had collected. The data were a part of a much bigger story, which was going to make a pretty big splash in a high-impact journal. I told them that I would not tell the complete research findings and scoop her big story, which she was preparing for. I said I would make this a side story. In the end, my research advisor said no, that I could not publish the work. I was disappointed but I also was set on graduating. I then went to my committee and explained the situation. After some conversations and explaining I was allowed to defend without the publication.

Despite the success of achieving my research committee's approval to graduate without a publication, writing the dissertation was really hard. I had trouble finding the motivation and did not feel my advisor's support. But ultimately, the yearning to graduate and moving onto my next venture got me through, and five and half years, I finally defended and finished the PhD. I was out.

Career transition: Over the course of my PhD, I had learned that I had many of the skills needed to be a data scientist. I was good at understanding data, conducting statistics, and interpreting results. I remember reading that data science would be the biggest field of the next few decades. Not only that, but the salaries were crazy high compared to what I had been paid during the PhD.

During the PhD, I continued to garner skills that would be helpful for my eventual career as a data scientist. As part of the PhD, I had to take elective courses, so to build skills in data science, I chose Intro to Computational Neuroscience and Intro to Deep Learning.

Before my PhD defense, I got into a data science boot camp and spent a few months in a data science boot camp before the dissertation and planned to do something after the dissertation.

When I tried to transition, I wanted anything related to data science. I applied to software engineering jobs, data science, and computer science positions, but did not get much response. In addition to looking for jobs, I applied to two data science bootcamps that help PhDs transition into data science: Insight Data Science Fellowship and the Data Incubator. I got into the Insight Data Science Fellowship and chose to go there. If I did not go through Insight Data Science Fellowship it would have been a lot more difficult. In that program, I was expected to have 90% of the skills of a data scientist and instead of teaching mostly, they make you do a project where you apply your skills. Eventually, the program sets up meetings with companies looking to hire data scientists. I presented my project to these companies. Unfortunately, the companies were not hiring at the time. When that happened, I had to keep on applying for jobs.

In January 2019, at about the time I finished the Insight Data Science Fellowship, I also graduated from my PhD. At this time, I connected with a recruiter on LinkedIn, and I got hired as a Data Science Coach at the Flatiron Data

Science Bootcamp in NYC. I moved to Manhattan, near where I grew up and where I currently work. After working at Flatiron School, I then got a job as a Data Scientist in a startup, where I worked for two years. At present, I switched to a home property intelligence company that provides data intelligence to insurance companies.

Learnings from academics: Overall, looking back, I do not think that I would have been a good fit for academics based on how I saw people who were good in academics. My research advisor was really successful. I believe her success was because she treated science like a business, and she made decisions based on what brought in the money. I also heard, from people who did postdocs, that their advisors said that you can work for them but you cannot take any ideas when you leave to start your own lab.

Throughout the five and half years of my PhD, I felt it difficult to connect with other academics. My understanding of academics had not been correct. I thought that the payoff for working longer hours, having more stress, and having lower pay in academics meant you were given the opportunity to pursue your passions and interests. That was not the case. Many academics, whom I met in my PhD, were focused on the profession of academia, rather than the excitement of science. For example, my advisor made decisions in her research lab, like switching from primates to mice as her animal model, based on what would get the most funding instead of being driven more by scientific rationale. What I learned was that the competition for grants and gaining esteem was so high-pressure and stressful that you had to do what would keep your career afloat, not necessarily what you were always passionate about.

I look back at that point in the second year of my PhD when I had found another lab to go into where I thought I could be happy. Even though I did not like the project, I should have probably switched out of that lab. After all, I had not intended on being in my research advisor's lab. I had wanted to be in her husband's lab. I was more comfortable with the new lab group and advisor. My takeaway from this was that I think it is more important to have an advisor that you get along with than just a project that you like because the advisor really sets the tone for your experience. I think I still would have ended up a data scientist, but I would have been happier during the PhD.

Career satisfaction: I am pretty satisfied with my career now. I like growing my career. When I was in academics it felt like there was only one path of being a professor and there were not enough careers available. Now there are a lot of careers. And I can focus on my technical skills and move up the individual contributor ladder or develop my people skills and become a data science manager. I like that I am learning a lot. What I really like about science is learning things and learning things about the world. And I feel like I am learning things about the world and I am still doing that in my current job. I am in a much better, happier place. My coworkers are happier than the ones in academics.

Advice: I would tell a person pursuing a PhD that there are a lot more options than there are in the past. You can continue in the academic path and be a professor and in industry there are so many different positions that will value someone like you. You can go into something technical or something that works with more people or if you are into tech like me then a lot of tech companies want to hire PhDs. And you do not have to just do tech, you can apply your PhD in something old fashioned like oil and gas or pharmaceutical drugs.

I went into neuroscience personally because I wanted to be challenged and be on the cusp of human thinking. Working as a data scientist I feel at home. There are always new skills, new areas to apply them to, and new problems to solve. Data science is a new field, so being in it I am on the leading edge of the field's development. For me, this is very fulfilling. It completes the desire I always have to learn more. It is what I am good at. That is the reason why I feel like I am in a better place.

Key Insights:

- It is important to choose a PhD advisor who you enjoy working with and for and who is someone you can trust.
- PhD groups with senior lab members or permanent research technicians can be a sign of a healthy lab to cultivate knowledge transfer.
- Using PhD courses can help create momentum for a future industry career.
- Taking the opportunities that come from a boot camp can help aid your career transition.

Transition Story: Giannis Gidaris, PhD

PhD: University of Notre Dame
Field of study: Earthquake Engineering
Current career position: Property Treaty Underwriter in Reinsurance

Motivation for pursuing a PhD: I was in Greece finishing my Master's degree in Earthquake Engineering at Aristotle University of Thessaloniki, when I decided to go for a PhD. I was interested in the research field I was in, and I thought I would like to continue doing more in a PhD. Back then, I considered following the academic professor route, but at that time it was mostly just a thought. For me, growing up in Greece, doing the PhD in the United States was kind of the dream and sounded very exciting to me.

I did not apply for different schools. Instead, my Master's advisor introduced me to my PhD advisor. My PhD advisor had been looking for PhD students and had reached out to my Master's advisor to see if my advisor knew any interested students. My Master's advisor in Greece recommended me. Then, my PhD advisor met me in Greece and explained the program to me and told me to apply and come to his group.

Advisor: My PhD advisor was a really great mentor. My advisor was always available, hands-on, and helped mold his students into good researchers and problem-solvers. My advisor was great at mentorship and was also a great teacher in the classes I took with him.

Environment: I did my PhD in five years. At the peak of my PhD advisor's group, there were six or seven people. At its smallest, it was three or four people. I think the small group allowed for my advisor to give us an ample amount of attention. My advisor was also good at fostering collaboration within the group. My advisor was great at managing the group in this way. I had heard that other advisors' encouraged sometimes unhealthy competition between students, but my advisor never did that. My advisor was good at helping and gave us our own projects so it was clear we were not competing with each other.

Project: The project I had was good. Initially, in the first year, I was figuring it out and my advisor made it clear that the goal of the first year was to primarily build a strong technical background and then begin research after the first year's classes in the summer. Overall, it was clear and my advisor steered and mentored me in the right direction.

Next directions: I did not start thinking of jobs until the fourth year of my PhD. At the beginning of the PhD, I started thinking I wanted to go into academia and of course, as I went on I started to see things more clearly. I figured I would try the academia path. I did not feel ready for faculty positions in my last year, so before the end of my PhD, I started approaching other professors' in the field for postdoc opportunities. At the same time, I applied for industry jobs and got one interview. Then, I had an offer from Rice University for a postdoc. I was not completely certain if I wanted academia or industry and that is how I decided to do the postdoc.

Postdoc: Before my postdoc, my priority was academics and then after the postdoc the prospect of working in industry started warming up for me. Then, at a certain point, it became clear to me to apply to industry. I got married and also wanted to find a new place where I could be with my wife, as she was evaluating various offers from Universities to do her PhD. These criteria were more important to me. Where I would go next would be to be with my wife, in the same place. When applying for industry jobs, I approached different people that were working in industry in different companies and asked how do you approach the job market. I got advice on how to tailor the CV and cover letter. Through my PhD and post-doctoral advisor, I found more jobs to apply to and got contacts to apply with and get jobs with. It was very friendly.

In my current job in industry where I recently changed roles within the same company from Natural Catastrophe Specialist to Reinsurance Treaty Underwriter, I am responsible for underwriting large and complex portfolios of risk written by primary insurance companies. I implement technical risk assessments, modeling, costing, and pricing of property reinsurance business. In my previous positions as a Natural Catastrophe Specialist, I was responsible for the development and improvement of catastrophe models for quantifying risk associated with natural hazards such as earthquakes and tropical cyclones. I started in a very technical field and all this knowledge from my PhD was good and useful and

I have learned new stuff. Of course, nothing is perfect, but in general, I am happy and I made the right choice.

Key learnings: A piece of advice that I would give to a PhD is to think and be honest with themselves. Consider do I like research enough? It is stressful. Do I mind its ups and downs? It can be lonely when you are the only one working on a project and even when you work on this when you are a professor you are alone. Consider, do I like working with my group mates, do I like the projects, am I okay doing that for the next three or four years. If there is no good relationship or chemistry, or good mentorship from your advisor, then it is a long time to just do it for whatever reason it is. Maybe you might consider switching projects or advisors but you have to make it good. The advisor plays a huge role. The relationship and mentorship with the advisor are important to your success. The culture of the group is very important. I had a great experience. However, I heard that when too much competition is encouraged it is not good and not healthy. It is important to see your group members as colleagues and collaborators and friends, if possible. I heard from other people that they made changes such as changing projects, research groups, etc., and then they were a lot happier.

Most importantly, I think it is important to assess and recalibrate. You might say the PhD is not for me and I will get a Master's degree and if that makes you happy then that is what you should do. But this was not the case for me. I saw from other people that sometimes things are not working and I saw how stressful and soul-crushing it could be. If you do not get excited, you will not get there as easily, and overall, you probably will not get there, in general, to your goal of academics anyway.

Advice for future PhDs: In hindsight, I think it would have been good to network earlier in the PhD. If you are determined to go into academia, it is important to try to build a network early on. Having this network can help the transition into professorship earlier. I think it is important to sow these connections. If one wants to go to industry, and I did not do this, but in hindsight, it would be good to do internships and industry sponsored graduate programs to see where you might want to work and to see what jobs are out there in industry.

Obtaining this knowledge about what career options are out there earlier, would help you steer your next move and tweak your plans. If you are interested in industry, you could adapt your coursework to be more industry-focused or to make you more attractive to your future employers. You could find out what courses, certifications, or training to get that would help you make the leap into industry earlier. In S and E, PhDs are in high demand in industry. And they are not the traditional jobs that PhDs would think of. There are PhDs needed in re(insurance), investment banking, hedge funds, consulting, etc. With these jobs, there are courses that one can take during the PhD to help you transition. Plus, if you were not sure about industry, taking a step to explore an industry career can really be helpful to deciding on industry or academics, and navigating into your future job.

Key Insights:

- Having a strong relationship with your advisor helps make the PhD more enjoyable.
- Checking in and being honest with yourself about what you want will help you decide on what you want to do with your career.
- Networking at conferences and within the academic community will help you want to stay in academics. By contrast, networking and taking proactive steps nonacademics will help you transition out of academics into your desired career.

Part IV

Becoming the Proactive PhD

13

Leveraging Your PhD

"Time waits for no one."

– The Rolling Stones

13.1 Importance of Using Your Time Wisely

It is no secret that time waits for no one. It is a long spoken phrase and mantra, like "carpe diem" (seize the day), and audentes fortuna juvat (fortune favors the bold). These Latin phrases and the long-spoken phrase by Geoffrey Chaucer, later perfected by The Rolling Stones, all reference the importance of utilizing your time wisely. Time is the common denominator of all people, no matter the socio-economic status or international origin. It is the fabric that binds us together. Yet, it is also the biggest responsibility of all.

13.2 How To Optimize Your PhD Year-By-Year

Time during the PhD and postdoc compounds as you progress and keep going. Because you have more knowledge and more time spent conducting research, you can be more productive. For example, the research in your postdoc should take less time than your first few experiments or projects. This is because you are more experienced. It is easy to be proactive during your PhD or postdoc and I have listed steps you can take in an example five-year PhD (Table 13.1). These tips and tricks can be extrapolated to whatever duration of your PhD. The career exploration steps can be used for postdocs making the transition to nonacademic careers.

How to Make Your PhD Work: A Guide for Creating a Career in Science and Engineering, First Edition. Thomas R. Coughlin.

Table 13.1 Example of proactive tips and tricks during an example five-year PhD.

Year	Exam	Proactive Tips and Tricks
1	*Qualifying exam*	*Research*
		• Qualifying exam: consider using your qualifying exam to write a review paper.
		• Learn as much as possible from your new setting and people.
		• Find mentors (both senior members of the department and professor level).
		• Become a strong researcher/learn what it takes to publish.
		• Explore external funding (e.g. NSF GRFP).
		Career exploration
		• Learn what others like about their PhDs and where they are going.
2		*Take diagnostics test #1*
		Research
		• Aim to start converting research into a publication.
		• Develop your research skills and learn techniques.
		Career exploration
		• Take courses on career development and earn certifications.
3	*Candidacy exam*	*Take diagnostics test #2*
		Research:
		• Design and conduct research.
		• Discuss and develop a plan for your research focus with your advisor.
		• Conduct your candidacy exam.
		Career exploration
		• Start to do self-assessment of careers.
		• Seek out internships in careers you find interesting.
		• Explore opportunities outside research (i.e. teaching opportunities, funding applications, and travel exchange programs).
4		*Diagnostics test #3*
		Research:
		• Continue research and update objectives.
		• Write up research results.
		• Submit research for publication.
		Career exploration
		• Formulate next steps.
		• Decide on nonacademic or academic career.
		• Attend grant writing workshops.

Table 13.1 (Continued)

Year	Exam	Proactive Tips and Tricks
5	*Dissertation defense*	*Research:* • Continue conducting research and writing to submit for publication. • Convert your publications and candidacy into a thesis. • Write your thesis. • Defend your thesis to your committee. *Career exploration* • Network more intentionally at conferences. • Seek out job referrals. • Apply for next steps.

In Table 13.1, this five-year outline depicts a PhD student using their first year to: turn their qualifying exam into a review paper, pursue grants to achieve financial freedom during their PhD, and learn from older PhDs to determine what options are available to PhDs who went through the same program. Then, in the second year, the PhD student takes the diagnostics test (Chapter 4) to determine how their PhD is going and to learn how to be proactive with their academic career. The PhD also begins research and starts to develop a research focus. For career exploration, a PhD begins to explore careers and seek out other opportunities available to learn more about careers.

The candidacy is typically conducted in the middle of the PhD or a year or two before the end of the PhD. By conducting the candidacy this PhD can begin to also formulate the written portion of this exam into a publication. Furthermore, the third year is a great time to explore careers and seek out internships in nonacademic careers. By taking a second diagnostics test the PhD will gain insights into their PhD to see if it has improved from the previous year.

In the fourth and fifth years, it is important to start to plan the next step. The results of the third diagnostics test in the fourth year provide an objective guide of how your PhD experience is going. In this five-year plan, it is important to utilize the fourth year to decide on nonacademic or academic careers. From here, informational interviews can be conducted and networking is increased. In the fourth year, it is important to start planning your publications if you have not done so already.

In the fifth year, it is critical that you wrap up your research, submit publications, and write your dissertation. The dissertation does not need to be scheduled more than three to four months in advance. Simultaneously, as you have defined your next steps with your career, as you are in your final year, networking and speaking with employers is key. Whether you are interviewing for a postdoc or looking for a position in industry, it is important to set up your postgraduation thesis plans before you defend.

Employers understand when PhDs do not have their degrees yet. It is common to write on your resume or CV, PhD, "expected" for the specific month and year of your planned defense.

13.3 PhD Defense

As the culmination of many years of hard work and dedication, preparing for a PhD defense can be both a daunting and exciting task (University of Rochester 2023). You might spend countless hours poring over your research and prepping your slides, feeling the pressure steadily building. But as nerve-wracking as the process may be, it is also a celebration of all the knowledge and expertise you have gained throughout your academic journey. Through thoughtful preparation and confident delivery, you have the ability to showcase the fruits of your labor and leave a lasting impression on your audience. With the right mindset, the defense becomes an opportunity to share your passion for your research, communicate your findings, and confidently answer any questions that come your way.

After you defend it is onward and upward for you to start your next steps. Helping for a few weeks to transition over information or planning to send files over email to your advisor are common final steps of a PhD.

13.4 Worst Case Scenario

Not taking proactive measures could leave PhDs and postdocs with less-than-ideal opportunities for their education level and expertise, leading to significantly less rewarding outcomes later on (Figure 13.1). Instead, it is important to recognize the benefits of initiating your own career search as soon as possible.

13.5 Layering Your Goals

Taking proactive steps toward a career with the same dedication and purpose as your PhD or postdoc is essential to ensure that you are not left wondering what your next step should be upon completion of your academic career. To do so, it is important to layer goals so that you can meet both career exploration and research obligations. This can be the difference between a frantic job search and a planned one. With layered goals of finishing your PhD or postdoc and conducting a

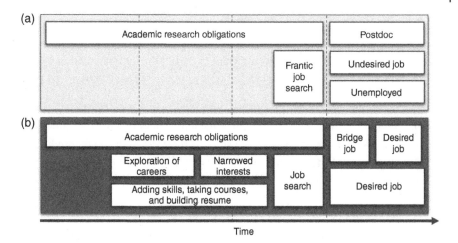

Figure 13.1 Beginning to take steps toward learning and exploring your PhD career options early will help you take more purposeful steps in your career. By balancing academic research obligations with proactive steps toward your career you will have a more focused and direct job transition. By contrast, having no plan will lead you to frantic job search and a job that you might not want.

purposeful job search, you can perform academic research obligations and take 30 minutes a day two to three times a week to explore careers and follow the plans in this book. This will give you the opportunity to identify and pursue potentially fulfilling careers that could play an important role in helping you reach your chosen objectives. And by making your future career aspirations a priority today, you will have the time required to turn such ambitions into reality.

13.6 Research Tips and Tricks

One of the main outcomes of your PhD experience is that you will likely improve your scientific and technical writing skills. PhDs become amazing writers by planning publications and working on grants.

Many PhDs have found that the below tips are helpful for optimizing their time during their PhD or postdoc. You will benefit from understanding these three grants and publication tips.

1) Before starting research layout the figures for your future research paper.
2) Gain PhD equivalence of financial independence by applying for grants.
3) Turn your exams into publications.

13.6.1 Start by Laying Out the Figures

Before even writing a research paper or beginning the experiment or analysis, sit down, and think about how you would want to lay out the figures. In other words, plan the figures that you would need in order to tell a complete story to get published ahead of time.

Most journal articles need greater than four figures in order to be deemed a worthy finding that has enough supporting evidence. In my experience, most articles use, on average, six figures to tell a story of their research. If you are having trouble deciding on the graphs, images, tables, or diagrams that would go in the prospective article, think about the tools and equipment that you have at your disposal. Consider a few questions:

- Which of these tools might you use to tell the story?
- What figure would each tool generate? A graph? Image? Table? Diagram?
- What is the one figure you would need to complete the story? What are the supporting figures that would answer the hypothesis?

Overall, laying out the figures allows you to decide which experiments you need to do in order to tell the story. This methodology can save months and sometimes years of dead ends. A wrong decision in research can equal months or even years of wasted time (Schreffler and Huecker 2023). By using this proactive method you will save time and energy, freeing up time for your continued development.

13.6.2 Gain PhD Equivalence of Financial Independence By Applying for Grants

There are a few PhD and postdoctoral fellowship-specific grants available. Obtaining your own funding allows you to be financially independent from your advisor and their salary. By doing this, you will have your own stipend. Having your own stipend, lets you work in a PhD advisor's group rather than working for the PhD advisor's group. That difference in freedom can allow you to be more independent, pursue your own interests, and develop your career.

For example, for PhD students, the National Science Foundation has the Graduate Research Fellowship Program (GRFP), which is available to first-year graduate students (National Science Foundation Graduate Research Fellowship Program 2022). This award provides students with a three-year annual stipend of $37,000 along with a $12,000 cost of education allowance for tuition and fees.

13.6.3 Turn Your PhD Exams Into Publications

The qualifying exam and candidacy can often feel like daunting tasks. Instead of spending hours upon hours writing documents that might not have any use to you or your advisor, consider exploring options like turning your qualifying exam into a review paper or using your candidacy exam as an opportunity to work on an actual publication you are hoping to write. Not only will this help you add valuable content to your CV, but it will also provide you with a sense of direction and purpose as you work toward your academic goals.

13.7 Preparing Your Career Early

As a PhD student, you are in the unique position of furthering your education while preparing for your career at the same time. It can be overwhelming to balance studying and job applications on top of other commitments, but these tips and tricks help make it possible to progress toward both goals.

1) Update your CV each semester
2) Keep track of your technical skills
3) Take career aptitude tests
4) Attend career training and networking events
5) Obtain training certifications
6) Find and apply for internships
7) Seek out mentorship and guidance
8) Stay up to date with technology and industry trends

13.7.1 Update Your CV Regularly

As a PhD student, updating your CV every semester may seem like a tedious task. However, it can greatly benefit your academic and professional pursuits. Not only does it serve as a record of your accomplishments and experiences, but it also allows you to reflect on your progress and identify areas for improvement. By keeping your CV up-to-date, you can easily tailor it to specific job or fellowship applications, showcase your academic and research abilities, and stay on top of your goals. So next time you find yourself putting off updating your CV, remember the many ways it can serve you in the long run as a PhD student.

13.7.2 Keep Track of Your Technical Skills

Gaining technical skills during your PhD or postdoc is a crucial aspect of your career development. These skills provide a foundation for your future career, whether it be in academia or industry. To ensure that you are keeping track of all the skills you gain, it is essential to document them properly. By doing so, you will be able to present a well-rounded portfolio of skills to potential employers or collaborators. It is important to keep in mind that these skills can range from programming languages to laboratory techniques. Therefore, make sure that you are not only noting the skills that are strictly related to your research but also those that can be applied more broadly. With this approach, you will be able to showcase your expertise while keeping your options open for the future.

13.7.3 Take Aptitude Tests

As you pursue your PhD or postdoc, it is important to not only focus on the academic and research aspects of your career but also to take the time to learn about yourself and your potential future career paths. Aptitude tests are a great resource for discovering your strengths and weaknesses, as well as identifying potential careers that would be a good fit for your unique skill set. Here are some the best ones:

- My Individual Development Plan (Science Careers 2003) – (www.myidp.sciencecareers.org)
- Myers-Briggs Type Indicator (16 Personalities 2023) – (www.myersbriggs.org/my-mbti-personality-type/mbti-basics)

By taking these tests during your graduate or postgraduate studies, you can gain valuable insights that will help you make informed decisions about your career trajectory. Do not be afraid to explore the many aptitude tests available to you – you never know what opportunities they may reveal!

13.7.4 Attend Career Training and Networking Events

As a PhD or postdoc student, career training and networking events hold immense importance in shaping the trajectory of your career. Attending such events provides you with an opportunity to meet industry professionals and experts in your field of study. Moreover, these events are tailored to provide you with valuable insights and training on how to identify and accomplish your career goals. It is essential to recognize that your academic research is not the only deciding factor in achieving career success.

13.7.5 Obtain Training Certifications

Attending career training and networking events during your PhD and postdoc provide opportunities for you to gain new insights, explore career options and connect with professionals in your field. They offer a chance to learn valuable skills such as public speaking, communication, and leadership. Not only can these events help you build a more comprehensive professional network, but they can also open doors to exciting job opportunities that you may not have considered otherwise.

By taking advantage of these opportunities, you may be able to take courses that you can add to your resume to help you with an eventual career transition.

13.7.6   Find and Apply for Internships

Conducting internships during your PhD and postdoc can be a game-changer for your career. The practical experience gained during internships can complement the theoretical knowledge learned in academia, leading to a more holistic approach to problem-solving. Moreover, internships can provide networking opportunities, exposure to industries and institutions, and increase your chances of securing a job post-graduation. It is important to carefully select internships that align with your research interests and career goals and to actively engage with your mentors and colleagues during your time there. With the fast-changing job market, internships have become more critical than ever for PhD and postdoc students to stand out in a competitive job market.

13.7.7   Seek Out Mentorship and Guidance Beyond Your Building

As you have come to know through this book, it is critical to gain career insights and perspectives and not just rely on your advisor's experience with PhD careers. Seeking mentorship and guidance from professionals outside of our department can introduce you to a wealth of diverse perspectives. Perhaps there is a leader in another department whose career trajectory you admire, or a team member in a different area of expertise who can offer valuable advice. Do not be afraid to step out of your comfort zone and seek out these opportunities. The knowledge and guidance gained from conversations with mentors can propel you forward in ways you cannot predict.

13.7.8   Stay Up to Date With Technology and Industry Trends

As a PhD or postdoc student, it is imperative to stay current with technology and industry trends in your field. Upon graduation, you will be entering a dynamic and ever-changing workforce, and having a deep understanding of the latest technology and market shifts will give you a competitive edge. Stay ahead of the curve by attending industry conferences, networking with professionals in your field, and keeping up with the latest industry news and reports. By staying informed and adaptable, you will be well-equipped to tackle challenges and opportunities as they arise in your career. Do not let complacency hinder your success – embrace the dynamic world of technology and industry trends and stay ahead of the game.

13.8 Conclusion

Overall, it is important to take the time needed during your PhD or postdoc to explore your career options and plan for the future. Taking proactive steps toward success can be difficult but it does not have to be. It is always better to plan ahead in order to maximize your productivity and knowledge acquisition. With research being a crucial part of the PhD experience, taking thoughtful measures now can lead to better results later on. Remember that each experience is unique, so these tips can be adapted according to what works for you. Ultimately, taking a few moments out of each day or week will go a long way in helping you meet the goals you set forth for yourself during this time period! Be sure to make use of any opportunities available and remember that there are many resources available if you need help along the way.

Chapter 13 Key Takeaways

❖ Staying proactive during your PhD will allow you to not waste time and make the most of the experience.

❖ Taking proactive steps in your research and career exploration throughout a PhD or postdoc is a vital way to progress through your degree into a job of your dreams.

❖ Research tips and tricks include: starting research by first laying out the figures, applying for grants to gain financial independence during your PhD or postdoc, and turning your PhD exams into written publications to not waste time during your degree.

❖ You can prepare early for your career during your PhD or postdoc by updating your CV regularly, keeping track of your technical skills, taking aptitude tests, attending career training and networking events, obtaining training certifications, finding and applying for internships, seeking out mentorship and guidance from mentors outside of your department, and staying up to date with technological and industry trends.

References

16 Personalities. (2023). Free personality test. NERIS Analytics Limited. http://www.16personalities.com/free-personality-test (accessed 11 June 2023).

National Science Foundation Graduate Research Fellowship Program. (2022). What is GRFP? www.nsfgrfp.org (accessed 11 June 2023).

Schreffler J and Huecker MR. (2023). Common pitfalls in the research process. Stat Pearls. https://www.ncbi.nlm.nih.gov/books/NBK568780/ (accessed 11 June 2023).

Science Careers. (2003). Home page. http://www.myidp.sciencecareers.org (accessed 27 April 2023).

University of Rochester. (2023). Preparing for a PhD defense. University of Rochester. https://www.rochester.edu/college/gradstudies/academics/phd-defense.html (accessed 11 June 2023).

14

The Future PhD

"A rising tide lifts all boats."

–John F. Kennedy

14.1 PhDs Are a Rare Breed

Only other PhDs and those very close to you will truly appreciate what you have experienced during your PhD. As PhDs, we have been through something together. And therefore, oftentimes you will find that many PhDs are willing to help you navigate your career. I suggest being open to making connections and to the warm welcomes and support you will receive. In fact, by contrast, you may be in a position to share your experiences and pay it forward. The PhD community needs to support one another. PhDs need to raise the tides to help support knowledge sharing and standardization of PhD career training.

As new PhDs find careers in today's workforce, there are many positive changes that have taken place across the PhD career landscape, including:

- Increased PhD support,
- More awareness of the need for PhD training,
- Increased resources for PhDs, and
- Movement toward artificial intelligence (AI) and remote work.

14.2 Increasing PhD Support

As you now know, a doctorate faces a drastically different employment landscape than at any other time in history. Despite the stark realization that PhDs are not just going into academic professor roles, the realization that there are a vast number of PhD careers in all sectors of the United States economy is, in fact, quite a welcome thought. To know that your PhD will hold value in other opportunities outside of academics, will help you to navigate the PhD. That said, the gap in understanding to get from academics to nonacademic careers centers around understanding what the difference is between these careers.

There are a number of trends that are impacting these problems and addressing this gap:

- There is more awareness by PhD graduate centers,
- There are an increasing amount of government programs promoting PhD career education,
- Nonprofit groups are forming, and
- Businesses have created training modules for nonacademic careers.

14.2.1 More Awareness of the Need for PhD Training

Change in academics is slow, and PhD job satisfaction is not a top priority and concern of many faculty members. Advocating for a change will need to come from an outside force and ideally a government program. There are some programs already enacting programs to promote careers after PhDs.

- **National Postdoctoral Association** – aims to improve the postdoctoral experience by supporting a culture of inclusive connection. At the individual, organizational, and national levels, they facilitate enhanced professional growth, raise awareness, and collaborate with stakeholders in the postdoctoral community (NPA 2023).
- **National Institutes of Health (NIH 2021) Broadening Experience in Scientific Training (BEST) Awards Program** – a two million dollar grant that promotes PhD and postdoctoral career training that extends beyond university campuses and into the for-profit industry, government, communication, and nonprofit corporations (BEST Program).
- **Graduate Career Offices and Individual Development Plans** – Many universities have expanded their career services offerings to include guidance for PhDs. They recommend individual development plans and sometimes have access to alumni networks for promoting jobs (myIDP 2003).

14.3 Increased Paid Resources for PhD Career Support

There are also other paid resources that have sprung up to fill in for the lack of education available within PhD programs to help them navigate career opportunities outside academics:

- SciPhD (http://SciPhD.com): Started in 2009, this program teaches a hands-on course to help PhDs switch from the academic way of thinking to the industry mindset (SciPhD 2009).
- **Cheeky Scientist LLC:** Started in 2012, this company focuses on creating content for helping PhDs understand the value of the PhD and delivers some free and paid ways to overcome the common plights of the PhD (Cheeky Scientist 2013).

14.4   Impact of COVID-19 and Artificial Intelligence

On 11 March 2020, COVID-19 was declared a pandemic by the World Health Organization. In the years that followed, the workforce and labor economy changed drastically. Academic institutions moved their courses online and professors taught their classes virtually using apps like Zoom, RingCentral, and Google Meet.

To slow the spread of COVID-19, many employers also shifted to remote work arrangements, which led to a significant increase in the number of people working from home. This has changed the way people work, with many companies realizing that remote work is a viable option and offering it as a permanent option.

In addition to increasing remote work, the pandemic has accelerated the adoption of automation and AI in the workplace as companies seek to reduce their dependence on human workers and increase efficiency. With the emergence of OpenAI's ChatGPT-3 (with ChatGPT-4 recently launched), the speed of companies offering AI solutions has greatly increased. The current market capitalization of AI companies was approximately 119.78 billion in 2022 and is expected to increase to 1.6 trillion by 2030.

14.5   PhDs Are Perfect for This New Work World

After the COVID-19 pandemic, there has been increased adoption of remote work and AI in companies. PhDs are prime for succeeding in this new career workforce in both academic and nonacademic. Oftentimes, what separates successful workers from unsuccessful ones is ability to adapt. PhDs are perfect workers to decide what technologies to integrate into their workplace. On top of that, PhDs can evaluate technologies to determine their value. These qualities will be crucial to companies in the coming years, as more and more companies attempt to keep up and adopt AI into their work technologies.

With the shift toward remote work, it is more important than ever to cultivate a sense of self-sufficiency and autonomy. Fortunately, PhDs are uniquely equipped to thrive in this new environment. Not only are you trained to think critically and make budget-conscious decisions, but you also have the ability to work independently without constant oversight. Whether it is developing a research project or managing a team of remote workers, PhDs are well-suited for the challenges of this new world. As a PhD you are rightly positioned to be productive in this new work world.

14.6 Conclusion

Despite some graduate career offices supporting career development and more paid resources for PhDs, there lacked a guide for how to navigate the PhD while addressing the pain point of telling you if you are competitive for a research faculty position. This book was designed for PhDs and postdoctoral fellows with both national and international backgrounds and it aimed to help you assess your PhD:

- using a program diagnostics test and
- stories of other PhDs who have been in similar situations, which
- led you to develop two paths that connected you to the right resources and the right career.

Chapter 14 Key Takeaways

- ❖ There are many positive changes happening to help support PhD career development.
- ❖ PhDs need to support each other through their degree programs.
- ❖ Many societies and nonprofits offer training programs for PhDs.
- ❖ PhDs are perfect for the remote world and resourceful individuals who can adapt to learn new technologies quickly.

References

Cheeky Scientist. (2013). About us. Cheeky Scientist. https://cheekyscientist.com/about-us (accessed 27 April 2023).

National Institute of Health. (2021). NIH common fund strengthening the biomedical research workforce. NIH Common Fund. https://commonfund.nih.gov/workforce (accessed 27 April 2023).

National Postdoctoral Association. (2023). About. National Postdoctoral Association. https://www.nationalpostdoc.org (accessed 27 March 2023).

Science Careers. (2003). Home Page. http://www.myidp.sciencecareers.org (accessed 27 April 2023).

SciPhD. (2009). About Us. SciPhD. https://sciphd.com (accessed 27 April 2023).

Appendix

Diagnostics Tests. The Diagnostics Test is discussed in Chapter 4 and can be used to diagnose and objectively assess how your PhD is going. We have three blank tests here for you to use during your PhD. See more details in Chapter 4 to qualify the results of your Diagnostics Tests.

Diagnostics Test 1

Date of Test: _____

1. Project (funding, project)

1a. Is your project a continuation of another project?	Yes (1)/No (0)
1b. Is your project something that your advisor or group needs to finish and move on from?	Yes (1)/No (0)
1c. Is your project new and novel?	Yes (0)/No (1)
1d. Is your project considered exciting in the field?	Yes (0)/No (1)
1e. Is your project dependent on someone else's work?	Yes (1)/No (0)
1f. Is your project on the cutting edge of the field?	Yes (0)/No (1)

2. Advisor (supervisor, mentorship, support)

2a. Does your advisor have weekly meetings with you or the group?	Yes (0)/No (1)
2b. Does your advisor give you advice on how to start research projects?	Yes (0)/No (1)
2c. Is your advisor encouraging?	Yes (0)/No (1)
2d. Does your advisor introduce you to people at meetings, conferences, or within your institution?	Yes (0)/No (1)

How to Make Your PhD Work: A Guide for Creating a Career in Science and Engineering, First Edition. Thomas R. Coughlin.
© 2024 John Wiley & Sons, Inc. Published 2024 by John Wiley & Sons, Inc.

2e. Does your advisor frequently check-in on you at your desk or in your lab?　　Yes (1)/No (0)

2f. Is your advisor present and there (within reason) when you need them?　　Yes (0)/No (1)

3. Environment (group, team, laboratory mates, institution)

3a. On average, are people in your group mostly available for help on a weekly basis?　　Yes (0)/No (1)

3b. Do people in your group share information freely?　　Yes (0)/No (1)

3c. Do the senior members in your advisor's group like coming into work?　　Yes (0)/No (1)

3d. For new PhDs, do the people in your group seem genuine?　　Yes (0)/No (1)

3e. Do other internal research groups compete with your group? Is the environment competitive?　　Yes (1)/No (0)

3f. Are the existing systems or protocols in place to make the work orderly and efficient?　　Yes (0)/No (1)

Tally your scores and write them below.

> **Total Scores:**
> 1. Project　　＿＿＿＿＿＿
> 2. Advisor　　＿＿＿＿＿＿
> 3. Environment　　＿＿＿＿＿＿

Scoring:

Good project, advisor, and/or environment	(score = 0–1)
Moderate project, advisor, and/or environment	(score = 2–3)
Poor project, advisor, and/or environment	(score = 4–5)
Terrible project, advisor, and/or environment	(score = 6)

Diagnostics Test 2

Date of Test: ＿＿＿＿＿＿

1. Project (funding, project)

1a. Is your project a continuation of another project?　　Yes (1)/No (0)

1b. Is your project something that your advisor or group needs to finish and move on from?　　Yes (1)/No (0)

1c. Is your project new and novel?	Yes (0)/No (1)
1d. Is your project considered exciting in the field?	Yes (0)/No (1)
1e. Is your project dependent on someone else's work?	Yes (1)/No (0)
1f. Is your project on the cutting edge of the field?	Yes (0)/No (1)

2. Advisor (supervisor, mentorship, support)

2a. Does your advisor have weekly meetings with you or the group?	Yes (0)/No (1)
2b. Does your advisor give you advice on how to start research projects?	Yes (0)/No (1)
2c. Is your advisor encouraging?	Yes (0)/No (1)
2d. Does your advisor introduce you to people at meetings, conferences, or within your institution?	Yes (0)/No (1)
2e. Does your advisor frequently check-in on you at your desk or in your lab?	Yes (1)/No (0)
2f. Is your advisor present and there (within reason) when you need them?	Yes (0)/No (1)

3. Environment (group, team, laboratory mates, institution)

3a. On average, are people in your group mostly available for help on a weekly basis?	Yes (0)/No (1)
3b. Do people in your group share information freely?	Yes (0)/No (1)
3c. Do the senior members in your advisor's group like coming into work?	Yes (0)/No (1)
3d. For new PhDs, do the people in your group seem genuine?	Yes (0)/No (1)
3e. Do other internal research groups compete with your group? Is the environment competitive?	Yes (1)/No (0)
3f. Are the existing systems or protocols in place to make the work orderly and efficient?	Yes (0)/No (1)

Tally your scores and write them below.

Total Scores:

1. Project   _____

2. Advisor   _____

3. Environment   _____

Scoring:

Good project, advisor, and/or environment	(score = 0–1)
Moderate project, advisor, and/or environment	(score = 2–3)
Poor project, advisor, and/or environment	(score = 4–5)
Terrible project, advisor, and/or environment	(score = 6)

Diagnostics Test 3

Date of Test: _____

1. Project (funding, project)

1a. Is your project a continuation of another project?	Yes (1)/No (0)
1b. Is your project something that your advisor or group needs to finish and move on from?	Yes (1)/No (0)
1c. Is your project new and novel?	Yes (0)/No (1)
1d. Is your project considered exciting in the field?	Yes (0)/No (1)
1e. Is your project dependent on someone else's work?	Yes (1)/No (0)
1f. Is your project on the cutting edge of the field?	Yes (0)/No (1)

2. Advisor (supervisor, mentorship, support)

2a. Does your advisor have weekly meetings with you or the group?	Yes (0)/No (1)
2b. Does your advisor give you advice on how to start research projects?	Yes (0)/No (1)
2c. Is your advisor encouraging?	Yes (0)/No (1)
2d. Does your advisor introduce you to people at meetings, conferences, or within your institution?	Yes (0)/No (1)
2e. Does your advisor frequently check-in on you at your desk or in your lab?	Yes (1)/No (0)
2f. Is your advisor present and there (within reason) when you need them?	Yes (0)/No (1)

3. Environment (group, team, laboratory mates, institution)

3a. On average, are people in your group mostly available for help on a weekly basis?	Yes (0)/No (1)

3b. Do people in your group share information freely?   Yes (0)/No (1)

3c. Do the senior members in your advisor's group like coming   Yes (0)/No (1)
into work?

3d. For new PhDs, do the people in your group seem genuine?   Yes (0)/No (1)

3e. Do other internal research groups compete with your   Yes (1)/No (0)
group? Is the environment competitive?

3f. Are the existing systems or protocols in place to make the   Yes (0)/No (1)
work orderly and efficient?

Tally your scores and write them below.

    Total Scores:

    1. Project   _____

    2. Advisor   _____

    3. Environment   _____

Scoring:

    Good project, advisor, and/or environment   (score = 0–1)

    Moderate project, advisor, and/or environment   (score = 2–3)

    Poor project, advisor, and/or environment   (score = 4–5)

    Terrible project, advisor, and/or environment   (score = 6)

Additional Resources

Professor Advice
- The Professor Is In, Karen Kelsky
- Next Gen PhD, Melanie V. Sinche

Self-Growth and Increasing Effectiveness
- 7 Habits of Highly Effective People, Stephen Covey
- 4 Hour Work Week, Tim Ferris
- The Intelligent Investor, Benjamin Graham

Career Readiness/Self-Assessments
- myidp.sciencecareers.org
- Myers-Briggs Type Indicator personality test – you can access a free one at: 16personalities.com
    - For building emotional intelligence and gaining perspective about workplace differences and building empathy for how people see and approach things differently

Articles on PhD and Postdoc
- nextscientist.com
- Specifically the article, "How To Know If You Should Leave Academia . . . Before Wasting Years In Postdocs"
- sciencemag.com

Free Postdoc to Industry Transition Advice
- stempeers.org
    - New network as of 2018 based out of New York. Cofounder Ananda Ghosh has visions of creating an international contingency of PhDs and postdocs who can share in their support of each other by posting stories, inspirational content, hosting web chats on important information, and helping others make the transition out of academics or decide to stay in

How to Make Your PhD Work: A Guide for Creating a Career in Science and Engineering, First Edition. Thomas R. Coughlin.

Single Cost Postdoc to Industry Transition Help/Network
- cheekyscientist.com
  - Very useful company that delivers high-value products and immediately hooks you into a network of postdocs that have transitioned to industry
  - Lots of shared experiences, informational interviews, and interview prep
  - Great value! Approximately $350

Transitioning to Pharma/Biotech
- biopharmadive.com
- fiercepharma.com
- fiercebiotech.com
  - Making the transition to industry pharmaceutical/biotechnology side, specifically for consulting, medical communications, or roles straddling business and science such as business development or tech transfer
  - These websites will give updates on any major newsbreaks in the pharmaceutical and biotech industry showing which drugs are doing well, when there are new clinical trials, important mergers or acquisitions of new technology, and much more
  - Gaining perspective on the industry can give yourself a gage if you are excited by the strategy side of new business, and understanding that money and economics drives the market more than anything else

Other Resources
- forbes.com
  - For self-management and management tips in workplace
  - Even dealing with workplace events or how to call in sick or how to craft informational interview questions

My Story

At the start of my PhD, my goal was to be an academic professor. When I thought long term, I envisioned myself at a small research- and teaching-focused institution living in the neighboring town, going home for lunch occasionally to let the dog out, and then jetting back to work to finish a lecture and grade. I was a little starry eyed.

Admittedly, the view of academics I had was a little romanticized, but I thought it could exist. There were professors at my PhD institution that seemed to have this balance of life and work. But they had all had tenure for many years.

Among new faculty, I saw that there is a work ethic that is required to make tenure for a research faculty position that compares to the discipline that it takes to be a top pro-level athlete. Personally, I remember not being scared of this. I thought I had developed good habits in my PhD to avoid burnout. I also felt like I had slightly outgrown the environment of my PhD and that a postdoc would likely be necessary. I did not have the research focus that would allow me to be an independent researcher. I needed a few more findings to get to that point.

Looking back, I felt an uncertainty in my career steps and realized in hindsight that I did not know all the options available to me at the end of my PhD. I felt that it was a gray space. And truthfully, I had been conditioned to think that academics was the only option and would be the only satisfying choice for someone as motivated as me. Oh, I was very, very wrong. And the lack of knowledge and the knowledge I would gain while navigating my career steps would lead me to start PhD Source, www.phdsource.com, and ultimately to write this book.

Finding a postdoc: I knew that I was not yet qualified for a research professor position, but I also had no idea what else was out there. Pursuing a postdoc felt like the logical next step. I received different advice and perspectives from many people about the postdoc. A tenured professor who had taught for twenty years encouraged me to advance my research career in a postdoc position, but then a newly hired professor told me only to do a postdoc if I was sure it could give me a new experience that could build on the research I had done. Knowing that the timing of that specific professor hiring me would have to coincide with my job search, I was not too hopeful. Plus, I wanted to come back to the East coast after my PhD in the Midwest. In total, I was able to secure four postdoc interviews.

In one interview at Princeton University, the student who gave me a tour of the lab of the principal investigator (PI) advised me not to come work at that lab. She told me that it was impossible to leave the lab, even to go to the restroom, without the advisor asking, "Hey, where are you going?" I immediately ruled out this option. I had heard enough stories of toxic labs after five years of my PhD to know to avoid this.

In the second interview, I met a professor who did not have much funding for my work but could afford to pay for me to be there in his lab. This professor told me that I would learn how to write grants during the postdoc. I felt interested in this position, but turned it down because there were no other postdocs and thus, I did not think the environment would be easy to navigate. In my third postdoc interview, I had a video conference interview with a professor in Boston. I was interested in the position, but there was not yet funding for the project and thus, no immediate employment. Finally, I went for an interview in New York City at NYU and found that the advisor was excited, kind, and had built a strong reputation for being a good researcher. He told me that an NIH postdoctoral fellowship with him would be almost guaranteed, and he would mentor me and teach me how to write the infamous postdoctoral NIH K99 grant that converts into an R01 level grant when you become a faculty position. He was a nontenured faculty member, which of course had inherent risks, but having my own funding to me quelled the risks associated with this. Truthfully, the lack of a top postdoc in a secure tenured faculty lab would eventually come back to haunt me.

My postdoc position: After a few months in NYU in New York, I started to have this unsettling feeling that things were not going well. The department was very toxic. The chair did not support my advisor's research.

Exploring nonacademic careers: Despite some of the problems, there was some silver lining; I was funded by the NIH and I had specific obligations with NYU to attend group meetings with other postdocs who had the same type of grant. I took business level courses about pharmaceutical drug development and gained perspective into the industry setting and realized that I could in fact find jobs that were interesting to me in industry. At first, I did not want an industry job. I associated an "industry" job with something uninteresting, controlled, with no creativity of utilizing my own ideas. But as time went on, my thoughts started to change. I attended, "What can you be with a PhD?" This event, held annually at NYU, hosts forums with PhDs who transitioned into nonacademic careers. In addition, I took courses hosted by SciPhD, courses in Team Science, and attended two of the Clinical Translational Science Institute (CTSI) meetings in Washington DC and Orthopedic Research Society Conference in San Diego, CA. From the CTSI meetings, I learned about careers in government, while I toured around the Senator's offices of New York and Mississippi to promote lobbying of increasing the NIH's budget. I was seeing the breadth of jobs outside of academics for PhDs who wanted to impact science and engineering, and not just do one job.

The final moment when I became convinced my career could fit outside academics was while taking a course called Biomedical Entrepreneurship at NYU. This course was led by very prominent entrepreneurs, including Ed Saltzman and Francois Nader, MD, MBA, who had led multiple drugs onto the market and described the trends and development of research in drug development. The course taught me two main things. First, I was, in fact, entrepreneurial-minded, and second, that an entrepreneur can do well in an industry position by using entrepreneurial skills. They called an entrepreneur in industry an "intrapreneur," a new idea of how to problem solve pain points and find opportunities while working at a company. This struck me as a great way of redefining a career objective. But I was not ready to call it quits on an academic tenure-track professor position just yet.

Postdoc game over: At the beginning of the second year of my postdoc, my advisor called me into his office to have a chat. I had turned things around with my postdoc at that point. I had published a review paper on biomedical research in post-traumatic osteoarthritis (PTOA) and decided to put aside the fluid flow research I was conducting and move into a research project that would yield interesting results regardless of the findings. It was a safe experiment with interesting implications unexpectedly. The smile quickly faded from my face as my advisor let me know that he did not receive the R01 grant that he had resubmitted for earlier that year. He said he was not invited to submit again, and his tenure review committee let him know he had to leave. He would be leaving for a new faculty position. He told me I could stay and use the funding from my postdoc to find a new position or move to the new lab in a different country.

At that time, I did not want to leave to go to a new country. I did not have any other job opportunities to jump into, and my healthcare, lease, and bills were being paid for by my postdoc grant. After my advisor left and one year remaining on my postdoc contract, I began reporting to the department chair, but he detested the work that I was doing. My personal life and feeling of progress began to crumble. To combat the feelings of insecurity and uncertainty, I leaned into the stable factors for my career and considered what I had going for me as options.

Next transition: I reached out to my network for the dreaded second postdoc, but my heart was no longer in research and the process of losing my progress, my postdoc, and the exhaustion of living on a shoestring budget in NYC with a year and half of work that had led nowhere was exhausting. It was not worth the risk.

I was concerned that another position might end in the same result or not get me close to an eventual secure position. I also disliked that I had seen a few PIs, including my postdoc advisor, and a few others during my PhD get denied tenure. I decided that I would be more secure in an industry job.

I started to apply to everything: data science, consulting jobs, professorships, teaching positions, medical writer positions, and engineering jobs. For six to nine months, I did not get a call back from any of them. I was shocked. I did not think it was possible for a PhD not to get job options. I spoke to other PhDs at NYU and learned my situation was not uncommon.

Finally, after nine months of finishing experiments, applying for jobs, and going to career networking events, I finally got a consulting position at NYU's Venture Capital fund. This was a bridge job and not a final position. But this position gave me something to list as job experience outside of academics on my resume, and I finally landed an interview at a medical communications agency in New York.

Nonacademic career: My career in industry since then has been full of many experiences. I was a part of an industry-wide layoff at this first medical communications agency I worked at; I started PhD Source, a website and blog with content for PhDs to help them in their career transition; and then I transitioned my career into startups and entrepreneurship. I started working as an adjunct professor of entrepreneurship and worked at a biotech startup and then fintech startup in New Jersey. These experiences finally led me to continue to put out content for PhDs with PhD Source, and then start my own company consulting for businesses and startups – writing business plans, coaching founders on pitches, and building pitch decks. Finally, I transitioned back into medical communications, where I am a senior medical director in pharmaceutical marketing and strategy.

Forging on: On a recent visit to my PhD's alma mater, I took a nightly run like I had done so many times during my PhD. I stood in front of my old research building and while looking at the building, I think I had expected to feel some of the old emotions I had felt during my PhD or perhaps the uncertainty of those PhD years. But instead, with the cool August air filling my lungs and the sound of crickets in the distance, I only felt a sense of calm pour through me. I felt a sense of pride for the decisions I had made and was proud of the person I had been and the person who I had transformed into. The feeling was fleeting and the responsibility of my present life reminded me of the distance that remained on the rest of my run. I started away from the building. Away from the old life I had, and felt the lightness and freedom of each step that I had felt all those years ago.

About the Author

Thomas R. Coughlin, PhD, was born in New Orleans, LA, and grew up in central New Jersey. He received his BS from The College of New Jersey and his PhD in Bioengineering from the University of Notre Dame. He has written 12 research journal publications, including one for *Nature Communications*, and presented at 23 conferences internationally. He has worked in consulting, taught entrepreneurship as a professor at Stevens Institute of Technology, and published papers on the state of the biotechnology sector in New York. He is currently living in Jersey City, NJ, where he works in pharmaceutical medical communications and helps PhDs navigate the complicated path from graduation to a fulfilling career.

How to Make Your PhD Work: A Guide for Creating a Career in Science and Engineering,
First Edition. Thomas R. Coughlin.
© 2024 John Wiley & Sons, Inc. Published 2024 by John Wiley & Sons, Inc.

202

Acknowledgments

Special thanks to my father, who is a writer and author of four fiction novels. If I never saw him have the hunger to create and his steadfast persistence, I would have never pursued this book, and likely the PhD, in the first place. My mother for all the phone calls she made during my PhD and postdoc as she took a strong interest in understanding my research and the academic world. She even helped out in the lab with me on bone harvest day in the second year of my PhD when she was there for Easter. My sister for giving me the courage to be myself and pursue my dreams. To Daniela for all the patience, encouragement, support, and creative discussions about the book. For all of my friends, who encouraged me to be who I am and who made the process of academics and getting a job all the more worthwhile. To my entrepreneur community at Stevens Institute of Technology, Matt Wade, Pat Sigmon, Sally Reiter, and Erike Mayo, your friendship has meant a lot to me.

My PhD advisor, Dr. Glen Niebur, for his amazing mentorship, writing critiques, the pints, the good times, jokes, and encouragement. And of course, my TCNJ advisor and mentor, Dr. Manish Paliwal, for encouraging me and seeing my potential before I did and for saying those special words that changed my life during my undergraduate, "You got accepted to Notre Dame for a PhD? Well, that's an offer you can't refuse."

When I started writing this book, my requests for interviews were warmly received and I am thankful for the PhD community who reaffirmed to me the importance of this book. Special thanks to my friends and colleagues – Laura, Henry, Giannis, Jesminara, Henry, Amar, Leon, Ada, Antonio, Sree, Vineeta, John, and Randy with SciPhD. This book would not be the same without you.

How to Make Your PhD Work: A Guide for Creating a Career in Science and Engineering,
First Edition. Thomas R. Coughlin.
© 2024 John Wiley & Sons, Inc. Published 2024 by John Wiley & Sons, Inc.

Index

How to Make Your PhD Work: A Guide for Creating a Career in Science and Engineering,
First Edition. Thomas R. Coughlin.
© 2024 John Wiley & Sons, Inc. Published 2024 by John Wiley & Sons, Inc.